SCIENCE A LA CARTE

And the Cherry Picking . . .
(fill in the dots)

A commentary and critique on current energy matters and their
relationships to man-made climate change

**Written by
Mathijs Beckers**

"The education of young people in science is at least as important, maybe more so, than the research itself."

— Glenn T. Seaborg —

"*Beware the irrational, however seductive. Shun the 'transcendent' and all who invite you to subordinate or annihilate yourself. Distrust compassion; prefer dignity for yourself and others. Don't be afraid to be thought arrogant or selfish. Picture all experts as if they were mammals. Never be a spectator of unfairness or stupidity. Seek out argument and disputation for their own sake; the grave will supply plenty of time for silence. Suspect your own motives, and all excuses. Do not live for others any more than you would expect others to live for you.*"

— Christopher Hitchens, Letters to a Young Contrarian —

This book is in **metric**

Numbers are in **short scale**

Foreword

The stakes: the continued survival of humanity and the prolongation of our scientific treks into the frontiers of the unknown, to keep pushing forward and to make up our own meaning, as none has been given to us.

Our world requires checks and balances, the press should be holding people accountable but instead media and journalists—with the exception of a few—are the messengers of those who pay them. I don't like this one bit. In order to make good decisions, regular people like you and me need to be well-informed. That's why I started to write books, to inform readers about the world around us; about what is happening and what other people are doing [*wrong*].

This book is meant to make you think, to nudge you into the direction of becoming more skeptical and also more knowledgeable about our world. What can we learn? What should we do in order to become more in tune with our natural surroundings, on Earth as well as in space? In order to make the right choices, we have to look at the ideas that are presented to us and weigh their value. Keep in mind that it should not be authority that determines what is true and what is not, but evidence and the way it is presented.

I love Cosmology and Biology, I am also a tech-head and a science-fiction and space fantasy fan, and as such I am constantly looking around for new and amazing things. The universe offers us a treasure chest filled with undiscovered wonders.

The fact that I am in love with science doesn't mean that I place scientists on a perpetual pedestal of reverence and awe, not unless they've earned it such as the likes of Archimedes, Pythagoras, Newton, Galileo, and Darwin. Even if their hypotheses, theories, laws, and principles might be disproved in the future. Their contributions are essential because they provide the cornerstones upon which we can expand our knowledge of the world.

Grudgingly I admit that there are scientists who present false arguments to advance ideological or even economic agendas. I invest most of my time in looking at the energy problems of our age, how these tie into the climate and what their effects on it are. During these studies I have found a breeding ground for fallacious articles, arguments and research—also take note that some of these make it through the peer-review progress. This is something we need to subject to careful scrutiny and if need be, point out as false or even dangerous.

It is very sad to see that highly estimable scientists and academics have joined the ranks of the nonsense-peddlers and this really bugs me! For I am a fan of some of these individuals, but fandom does not equal boundless restraint of criticism.

It is paramount to acknowledge that there are individuals that claim that what they do is tremendously difficult and as such they will conjure up strange yet meaningless metrics and try to obfuscate things in order to make their own cases stick. They do not stick, at least not in a scientific sense anyway, because there are enough skeptical people around to question their ideas and hypotheses. With this book, I intend to show you how easily some of these people can hoodwink you into believing stuff that simply isn't true. These claims are counterfactual—what might be, if...

Let's push ahead and have a critical look at what we can learn from all of this. Remain skeptical, keep demanding evidence and question everything. Only through these principles may we advance into a future state in which we will leave this world a better place than we found it.

Note, this book contains contrarian views regarding renewable- and nuclear energy.

Be forewarned

I do not shun the profane, I am a freedom of speech absolutist, this book might offend you because I will be pushing the envelope in this one.

"I am I hope never offensive by accident"

— Christopher Hitchens —

This book is in metric unless stated otherwise. This book contains contrarian views. I am not an authority, nor an expert. This book contains terse critiques, consider it a commentary.

Use your mind

Fair use statement

This book may contain copyrighted material the use of which has not always been specifically authorized by the copyright owner. I make such material available in an effort to advance understanding of environmental, political, economic, and scientific issues.

Also, some quotes by published Authors and Scientists are [mostly] meant as a compliment, I couldn't have written it any better than they have. It would please me tremendously if you, my dear reader, would buy the books from which these quotes came because they are of immense quality and will enrich your view on our world tremendously.

This book has been written specifically to address statements made by individuals from all over the globe and to offer help or criticism or an alternative explanation.

Cover design by
Robyn Gough

Ocean change & arctic methane

I am part of a relatively new yet unnamed and non-corporeal movement. This is a movement of people that understand [parts of] the sciences behind anthropogenic climate change and get up in the morning with a dedication to fixing the ills of our age.

The purpose of this chapter is to briefly explain the destruction we have wrought on the planet, the biosphere and to plant a sense of urgency in order to make sense of what is going to follow while we progress in this joint trek through this book.

The basics are these, the energy that is stored in the atmosphere, the oceans and in the ground predominantly comes from the sun. The sun irradiates the Earth, some of this energy from the sun gets reflected back into space—by clouds, ice and snow i.e. white and light-colored, smooth surfaces—and a lot of it gets absorbed in the atmosphere and the surface area of the earth. A dark surface warms up because it absorbs sunlight (energy) and subsequently turns it into heat (infrared) and radiates it back into the air towards space, and greenhouse gases in the air trap some of this heat, warming the air molecules around them. This is a big reason why our planet is habitable in the first place because without these gases our planet would be stone-cold.

There are two reasons why more heat would get trapped in the atmosphere: First, the increase of darker surface area due to ice- and snow melt—as is happening in the arctic and on the mountains; Second, an increase of important greenhouse gas concentrations in

1

the atmosphere—mainly water vapor, methane and carbon dioxide. Retroactively we would see a cooling of the earth if there would be more reflection and less heat trapped in the atmosphere through lower concentrations of different greenhouse gases.

An excellent explanatory diagram is provided by NASA:

http://science-edu.larc.nasa.gov/energy_budget/pdf/ERB-poster-combined-update-3.2014.pdf

We have caused the temperature to rise by emitting massive amounts of more potent greenhouse gases such as carbon dioxide and methane. If you look at the composition of the atmosphere, we can however barely see the volume of potent greenhouse gases, and one starts to wonder how these could have such an influence on the heat-trapping capacity of the atmosphere. For this, I have to redirect you to a book written by Norwegian scientist Svante Arrhenius in 1906 called *Worlds in the Making, the evolution of the universe.* This book is freely available in digital form on the internet; you will find a link in the appendix. Read this short excerpt:

"Any doubling of the percentage of carbon dioxide in the air would raise the temperature of the earth's surface by 4 degrees; and if the carbon dioxide were increased fourfold, the temperature would rise by 8 degrees."

110 years ago a man could already make these predictions based on observations from that age... The fact that Carbon Dioxide is just 400 Parts per Million does not negate its potency. In fact, it is a testament to how effective these molecules are at absorbing heat and transferring this heat to their neighboring molecules. The increased energy transfer from these carbon dioxide molecules then ripples through the air; you may liken it as a wave effect caused by a drop.

Science a la carte

Climate and Ocean Change are real. I won't accept the *controversy* that is being played upon by many people that have been deluded or have deluded themselves into walking—seemingly with their heads in the clouds—into oblivion. The debate should be over... There's no such controversy. We hear arguments like the climate has been changing for millions of years, it's the sun, it's subsea volcanoes, and so on but we've already had conclusive evidence for decades that it is we who have worsened and disrupted the amplitudes in climate and ocean change. We have altered it by changing the chemical composition of the atmosphere and the oceans. And we've changed the climate of Earth in such a way that it will bring about civil and geological and natural instabilities, famine and possibly large extinctions.

A great and enlightening video on this subject is called "*Dr. Richard Milne - Critical Thinking on Climate Change: Separating Skepticism from Denial.*" which can be found on YouTube.

If you are scientifically literate I suggest you examine this article (a link to the PDF below) shared by James Hansen and a significant amount of scientists including:

Makiko Sato, Paul Hearty, Reto Ruedy, Maxwell Kelley, Valerie Masson-Delmotte, Gary Russell, George Tselioudis, Junji Cao, Eric Rignot, Isabella Velicogna, Blair Tormey, Bailey Donovan, Evgeniya Kandiano, Karina von Schuckmann, Pushker Kharecha, Allegra N. Legrande, Michael Bauer, and Kwok-Wai Lo—credit where credit is due.

"*Ice melt, sea level rise and superstorms: evidence from paleoclimate data, climate modeling, and modern observations that 2 °C global warming could be dangerous*"

http://www.atmos-chem-phys.net/16/3761/2016/acp-16-3761-2016.pdf

Also, consider Dr. Michael E. Mann's (Author of "*The Hockey Stick and the Climate Wars*") explanation which he gave on the Rubin Report in March 2016. I Paraphrase slightly here for clarity:

"*what we're doing is we're taking all of that carbon that was buried naturally by the earth over that ensuing hundred million years since the Early Cretaceous as the climate slowly cooled down nature slowly buried all that carbon. We're putting it back into the atmosphere and we're doing it over a time scale of about a century, a **million times faster** than what nature would be able to do on its own and so that's really the issue we're talking about rates of change that exceed anything.*"

Note: that Dr. Mann is enamored by Mark Z. Jacobson 100% WWS scheme, which I am about to critique. WWS is short for wind, water, and solar energy. This is something that bothers me. Bill Nye is also enamored by this idea. Come on Guys! You inspired me to be skeptical and do my own research and start writing these books. Get real!!!

It is quite clear, isn't it? We know what greenhouse gases do, we have known this for ages. We also know that there have been fluctuations in carbon dioxide concentrations over the many ages of the Earth, but we have never seen such a steep incline in carbon dioxide concentrations up until the industrial revolution, the age of combusting fossil fuels.

It is not the stature or expertise or credibility that merits one to be taken seriously, it is the volume of evidence presented to substantiate their claims. How is this evidence gathered? It is gathered from all over the world by scientists making observations using all sorts of different sensors both underwater, on the surface of the water, on land, in the ice, in the air, and from space. It's the collection of this data, the corroboration thereof and the interpretation thereof which are based in physics, mathematics, statistics and chemistry that help us understand what is happening to our planet and determine cause and effect.

Life arose from the oceans. Billions of years ago our ancestors crawled from the sea onto the land. Over a long period of time—through the process known to be evolution—over many generations these creatures adapted from aquatic to becoming semi-aquatic to becoming creatures that lived on the land and in the air. The ocean is the cradle of life on Earth and remains so, even today for it is where the base of the food pyramid is. And, it's an important source of what all animals need to live: oxygen.

A fellow environmentalist and friend of mine—Doctor Alex Cannara—asked me to change the name of this chapter, instead of using Ocean Acidification, he convinced me to call it Ocean Change. Why call it Ocean Change, you'd ask? Because the oceans are not only acidifying, they are rising and being polluted, they are becoming warmer, dead zones are growing due to oxygen depletion and the changing of the direction of nutrient rich currents, El-Niño's are becoming stronger, storms are becoming stronger in some regions whereas other areas are being deprived of precipitation and we may note that there are many more effects that have been wrought on our Earth's oceans.

Where it is true that sea-level rise will cause billions of people to become displaced and trillions worth of ocean-side real estate to become worthless, it is a matter that will take quite a while to become a reality—in the meantime, we've got even more dangerous matters to attend to—who would have thought...

Consider for instance the current refugee crises all over the world, not just from the middle-east, but also from Africa and parts of Asia. Civil unrest and strife are often exacerbated by shortages of water and food and medicine and energy. In fact, it is believed by researchers that the Arab Spring has spread the way it has spread

5

because of crop failures in Syria and Russia, which mainly is being exported to Arab and North African countries. Hunger exacerbates civil unrest because it impacts human dignity and pride. The leaders that were overthrown in these countries sat on their thrones for quite a while, and one of them is still seated there, regardless of the destruction that war and terrorism without end have wrought on his country—Syria.

Also, note that fresh water supplies all over the world are being strained. The issue with one of the changing processes tied to the ocean—the hydrological cycle—is that it will change the precipitation patterns that are required to replenish water supplies like snowpacks, glaciers, rivers, lakes and aquifers. If the hydrological cycle gets altered, like it is in the Amazon, for instance, one can observe ever more drying patches of land due to less precipitation, or unbalanced precipitation. Rather than getting a constant, yet manageable influx of fresh water, you either get torrential rains or no rains at all. The issue with torrential rains is that less of it gets absorbed into the soil. Most of it will run off while it also takes with it parts of nutrient rich soil while it also might cause major flooding and other dangerous events.

These changes in the hydrological cycle not only start with an ever-heating ocean but also with the loss of tree cover which leads to a decline in evapotranspiration. What is evapotranspiration? Plants and trees pump up water from the soil and use it for photosynthesis, the water that has been pumped up, that is too much, is transpired into the air. It's not at all that unlike human transpiration. Evapotranspiration is one of the reasons why there is so much rain in the rainforest of South America. The other reason there is so much rain is due to the seeding of clouds by introducing small particulates on which water vapor can condense, these are called cloud condensation nuclei.

Science a la carte

Think about the food baskets of the world, what will happen as soon as water replenishment rates drop below draw-rates? I mean water drawn for anthropogenic activities such as farming and production and sanitation. If you consider some of the maps shared by the WHO and other organizations you clearly see that many areas already have large water troubles, and it is predicted that these are going to get more severe. If we run out of fresh water in region X, we're done there. There's only so much you can do without fresh-water. Even worse if this water scarcity hits places of high importance to our food supply we will see crop-failures and with ensuing shortages civil unrest will remain just around the corner and ongoing.

Where it is true that some areas have become reforested since the start of the industrial age, the net amount of natural forest cover in the world is a showing a decline. Especially in areas that are well-suited for crop cultivation such as the tropical rainforests around the Equator. Vast areas in South-East Asia and Africa and South America are being clear-cut for a variety of reasons, all of which are tied directly to human greed but also due to the need to survive, with which I can sympathize.

I must admit that I have simplified these issues, but I do this for a reason because I want you to understand what is going on in the world. Through this increased understanding I hope I can help you understand the things I am going to tell you in the coming chapters by planting the seeds of urgency, engaging the logical parts of your mind, and asking you to take nothing at face value, but examine the evidence for yourself.

With the increased storage of energy from the sun in the oceans, another periodical process of the hydrological cycle gets more intense: El-Nino. When an El-Nino occurs, warmer water rises to the surface in the Pacific Ocean. These increased temperatures help form stronger storm systems and increase short yet massive bursts of precipitation. This could mean alleviation of water issues, but also worsen them. Consider for instance the poor water management networks of the Asian Subcontinent. With torrential rains, the runoff increases and with it also the risk of flooding and crop losses. Or look at California's elaborate system of water canals, and drains and other facilities, an increase in precipitation will alleviate their water issues, albeit temporarily. The Sierra Nevada is one of the most important regions to hold fresh water supplies for California, and the snowpack has been declining for years. During the 2015-2016 El-Nino it finally got replenished to some degree after a long and steady run of decline. And so the El-Nino has brought about some—albeit very little—relief to California and severe troubles elsewhere. It gives one pause that a process that has been worsened by man-made climate change can bring some alleviation of any kind, but one has also to consider the ramifications in the long run. The net effects are believed to be negative rather than positive.

Also, consider the effects of melting glaciers and snow packs in the Northern Atlantic, thanks to the increased cooling effect (which sounds counterintuitive after all I've said) may disrupt the oceanic gyres as described in the article by James Hansen et. al shared with you earlier. The effects on the Atlantic Ocean could be profound, it will probably catalyze and produce more severe weather patterns all over the world.

The final and most dangerous development of all is the possibility of a subsea blight thanks to the acidification of the oceans through our increased carbon emissions. If we continue putting carbon dioxide into the atmosphere, we risk lowering the pH of the oceans to such a degree that plankton and coral reefs start dying out. When this happen, we've set a long string of dominos in motion, and there is no telling where the stones will stop to drop. In my previous book, I explained how the pH of the oceans is being lowered by increased carbon dioxide in the atmosphere; I will re-iterate it albeit trying to do so more efficiently, as I will learn from it myself.

At the surface of the ocean different gases go into and come out of solution. One of these gases is carbon dioxide which is known as CO_2. When CO_2 gets absorbed by water (H_2O) it chemically bonds with it creating HCO_3^- and one H^+ ion, subsequently another H^+ ion gets released and we're left with CO_3^{2-} and an additional H^+ ion. The amount of H^+ ions per volume of water determines the acidity. If there are more H^+ ions the water is more acidic. It is thought that a pH of 8.1 / 8.0 is the critical threshold at which important oceanic organisms such as Plankton start to dissolve and may go extinct.

Why are these organisms so important?

Plankton provide oxygen and food, through photosynthesis, for practically all of marine life. Through photosynthesis, they grow and gain more mass. It is their way of existing within nature. In term sardines and lots of bigger filter-feeding fish depend on the existence of plankton for their own nourishment. If plankton goes, so does the rest of marine life. It will also disrupt the carbon cycle even more as less carbon dioxide gets absorbed and turned into biomass. We have to keep a keen eye on the oceans as they are the storehouse for many organisms vital to the biosphere.

Beside these issues, anthropogenic climate change is causing extinctions throughout the biological realm. Once species are gone, they are gone. We are losing biodiversity! The beauty of nature is slipping through our hands. It will take millions of years of speciation and natural selection for new species to arise if it even can happen because we are really messing this place up with massive amounts of pollution and the chemical alteration of the oceans and the atmosphere.

The same issues arise with the increased heat content in the oceans, it disrupts life as it is and it contributes to the bleaching of coral reefs. Also, it contributes to the expansion of dead zones, it pushes nutrient rich waters into areas where there's less sunlight available for life-permitting processes, and so on and so forth.

The final issue is in dispute, not whether it will happen, but to which extent and what influence it will have on the bigger scale of things. As the ice cover of the arctic declines over the summer, it exposes both darker landmass and the sea. This means that more energy gets absorbed by the arctic, which subsequently leads to again less ice to be formed, and once again more heat absorption. It is hypothesized that a warmer arctic may influence or disrupt oceanic circulation patterns.

It is now well known that the arctic is a store of methane, which is sequestered in subsea permafrost and in the permafrost of the Tundra. If the arctic heats too much, it might give the earth a potent kick through the increased yet brief release of vast quantities of methane into the atmosphere, and this might increase temperatures further.

Again I have to admit that these matters are far more complex than the simplistic picture I am painting here, but it serves to drive home the message that we're up against a formidable foe and a couple of dangerous henchmen. We have to stop the destruction of vital parts of the hydrological cycle and stop the emission of dangerous and harmful emissions or we might not make it.

Please consider the studies of Universities, NASA, and the IPCC. Also, consider the books and papers and talks shared by countless of scientists and academics like James Hansen, Ken Caldeira, Tom Wigley, Kerry Emmanuel, Barry Brook, George Monbiot, Stephen Tindale, and many more. I know a couple of these people and they—like me—are true and very concerned environmentalists. These people want to fix civilization, and its vexing influence on our planet, through careful investigation and progress and innovation and the maximization of human potential everywhere on planet Earth and beyond.

We have to pump the brakes now, once positive feedbacks kick in, we can do whatever we want, and it isn't going to matter.

I will leave some links in the Appendix in order for you to be able to start your journey on becoming more knowledgeable about these issues if you haven't already. Remember, we can only make correct and effective choices if we are well-informed—and I am confident that the battle against ignorance will be won, but we're not there yet. You and I have to keep this ball rolling.

Solving this conundrum is not only a technological issue but also an economic, political and even religious issue - even though many will not admit it. I will continue writing this book focusing on the technological issues that worsen man-made climate change and those that may eventually solve it.

"Democracy cannot succeed unless those who express their choice are prepared to choose wisely. The real safeguard of democracy, therefore, is education."—Franklin Delano Roosevelt

Misnomers and non-solutions

We need to do something about the way we create energy because that determines a great deal our influence on the climate and the way it will change. We are making climate change worse by emitting carbon dioxide and other potent greenhouse gases.

Since the dawn of the steam-engine, we have developed a toolkit for energy generation. A couple of technologies have become mainstream, and these are coal-fired power plants, gas-fired power plants, biomass-fired power plants, diesel generators, nuclear power plants, dams, wind turbines and solar panels. In this chapter, I will mainly focus on the misnamed technologies, the ones that are considered to be renewable. There's a vast and pervasive cloud of misinformation surrounding these technologies and in order to make correct decisions we need to take a closer look and try to get some clarity.

The photoelectric effect is one of the amazing and wondrous deductions of Einstein. We can create panels that capture certain photons that were radiated from the sun and turn them into electric current. It behaves much like a chemical reaction, not unlike photosynthesis. You may simply consider it to be the exciting and exchanging of charged particles, which is the basis for chemistry.

"*All we need to do is capture the energy from that beautiful nuclear fusion reactor in the sky*" is the most heard argument together with "*More energy strikes the Earth in an hour than what we can use for a lifetime*" or some variations of it. It is all true, but we need to capture it first, and capturing energy—farming is a better word—requires billions of solar panels that do this job for us.

Almost everybody involved in selling PV panels tells the buyer that these panels are safe and clean. This is not true; in fact, it is a [white] lie. It might sound infantilizing, but there's no other way around it: PV panels are not created by unicorns and fairies, they are manufactured in large factories after their rough materials have been mined and refined and purified, it is a real industrial process. The scale of this is mind-boggling, we currently produce somewhere between two- and five hundred million PV panels per year, and all of these get transported around the world in large ships and tens of thousands of trucks. The idea that we can defeat BIG X with BIG Y is fallacious at best. PV is another big industry.

Bill Gates once said that PV doesn't destroy the planet (2011 WIRED Business Conference). I don't know whether he has since changed his mind, but if it still stands, I disagree. To substantiate this, I only have to tell you that an average PV panel weighs somewhere between 10 and 20 kilograms. High-quality quartz, copper, and rare earth materials are required to manufacture and build these panels, and we have to acknowledge that we would have to build billions of them. In addition, they wear out in a few decades, so we would also have to mine new materials, consume new energy and recycle (requiring more energy) billions of them annually. Plainly said this means that someone has to start shoveling and in the ensuing process many thousands of tons of toxic byproducts will be dumped in dystopian places like lake Baotou in Central Asia.

Also, consider the fact that current manufacturing capacities are capped due to the limited availability of the rare earth materials that are required to boost the chemistry of PV panels. Want to produce more? Again, start shoveling... PV isn't free, nor sustainable. These simple notions haven't arrived yet, that's all.

In the end, you may regard the solar panel, the one that works on the Photovoltaic principle to be just another gimmick. Solar Panels have not the potential to disrupt business as usual where fossil fuels are concerned. This technology has too many constraints in reliability, the need for constant offsetting, and the need for storage. Compounding issues such as the limitations on production capabilities, and resource limitations are set in stone and there's only so much you can do per square meter of energy.

I do think that PV could be an excellent bridging technology to move rural or remote communities forward towards a future in energy prosperity. Quite honestly, I wish for PV to remain niche because it's going to become too bothersome in terms of waste once we get into the realm of cumulative upkeep. And we have to get serious about efficiency, and not just the efficiency that people call "conservation", you cannot have your *conservation a la carte*.

The final pervasive myth that needs to be busted before we leave photovoltaics behind us is: "*Solar still may grow and new technologies will be invented, innovation is far from over.*"

If you download the *"Best Research-Cell Efficiencies"* from the NREL website (National Renewable Energy Laboratory), you will see a timeline from 1975 until 2015 in which is plotted the growth of efficiencies of different kinds of PV panels. You may note that at the start in 1976 the highest efficiency was 22% and in the end around 2013 44%. We may, therefore, conclude that over 37 years the maximum efficiency of solar panels grew by 0.6% per year. These are increments of 0.1% which are equal to roughly 1.4 watts per square meter. One may also note that for forty years people have been trying to perfect solar technology, and as such we may call it quite mature. Sure there are still different chemistries that may prove to be even more efficient but we cannot possibly maintain the position that PV is this fledgling technology that is going to skyrocket suddenly; that's far too simplistic and naive. If it truly had potential, it would have already taken off, despite decades of subsidies to boost deployment and make them profitable.

An efficiency of 44% yields about 600 Watt per square meter during optimal circumstances (highest solar exposure), and there is no conceivable way to increase the capacity factor significantly. Not even batteries are going to change these facts. If we are planning on adding batteries to this civilizational conundrum of ours, we would increase the volume of panels required again and again because we would need to create enough electricity to satiate demand as is, plus manufacturing the batteries, and account for future demand during solar-lulls and night-time usage. Besides this issue, adding storage requires adding more material requirements to an already ineffective technology thus making it even less preferable. We need to be doing more with less, not less with more.

One solar technology is left for me to brand near useless, which is known as CSP or Concentrated Solar Power. CSP is an Archimedes death ray aimed at a tower that holds a molten salt loop or a boiler. A vast array of mirrors is placed around a large tower, where all these mirrors focus solar rays to a single point, and this point is heated to several hundred degrees Celsius. The heated substance (usually molten salts or steam) is then the energy carrier that makes sure that a generator gets put to work, generating electricity.

The Ivanpah plant in California is one of the biggest and most well-known CSP plants in the world. It is not known for its success, however, but for its possible demise. Consider that this 2.2 billion dollar project of which 1.5 billion dollars has been invested on behalf of the people of the United States—government money. Remember this because it will become more relevant later on in this book. Ivanpah does not deliver as promised and in the process, it manages to scorch numerous [rare] birds.

Also, consider that Ivanpah requires natural gas to get started every day, and it uses much more gas than was initially expected. This is a very crude solar / gas fired power station at a disappointing nameplate capacity of 377MW and a limited capacity factor of 31%.

In 2016, Ben Heard—on the advisory board of Terrestrial Energy—wrote an interesting article on decarbonisesa.com in response to Ivanpah's imminent failure. He argues that it should remain operating in order to see whether the operators can fix the issues with which it is mired. Secondly, if it is not possible to get it running optimally, that it may serve as a reminder of what technologies not to implement. I think this is an intellectually honest position to hold, from an empirical standpoint it is the best possible way, as you get to observe it, analyze it and derive conclusions from it. It adds to the learning curve which may turn it into a success nonetheless, albeit not in terms of energy generation.

Is there then absolutely no hope for solar energy? The answer is no! While PV and Solar Storage (the concentrating mirror approach) aren't viable alternatives to base load energy, there is a technology that can be used on a great scale and which will alleviate energy use, Solar Heating. Many solar heating systems look like PV systems, but they are fundamentally different. We can run a liquid through dark tubes that are exposed to sun rays, which in turn transfer the solar energy that strikes the tubes into the liquid, which in turn transfers this heat into a big water tank / boiler. By using this method of transferring heat from the sun into your central heating system, we can alleviate the need for electricity or natural gas. The heat will dissipate slowly because we use a well-insulated container. This technology does not require any rare materials and is not constricted. In most countries central heating systems are being fed with gas or coal or electricity. Introducing the ability to capture latent heat will ensure that we can heat our water using less energy which makes this exception. Solar heating? Not a bad idea, albeit with some technical challenges down the line.

Another mature energy farming technology is wind. Wind contains kinetic energy that can be converted into a rotating motion through the use of turbine blades. Wind turbine blades are attached to a shaft, which drives a gearbox, which drives a generator that produces electricity. It's a fairly simple technology, of which simpler iterations have been in use for thousands of years now. The Dutch used windmills to pump water out of lakes in order to create new land, the first mention of these water-pumping-windmills dates back as early as the year 1316. This is as mature as it gets! The Dutch have been using mechanized processes to conquer the waters for nearly 700 years. What are you going to do when the sea threatens your shorelines? You call the Dutch!

I am fond of the idea of using mechanical wind power, to use for pumping or crushing, it is relatively simple to do and non-invasive! Yet it doesn't work all the time, and that's the clue.

Electricity from wind is a different story. The biggest wind-turbine we can build has a capacity of 8 MW, and it is massive! The total rotor blade diameter of this machine alone is 164 meters. A wind turbine needs a large block of concrete at its base, it has a steel structure, heavy turbine blades and a mechanical box that needs routine checkups and maintenance and they also often need heating during winter to keep the lubricant flowing and the electronics at the correct temperature. Fire hazards in wind turbines are real because they contain flammable gearbox fluids and other lubricants, and an electrical or mechanical failure may start a fire. Also, consider that material fatigue is a real thing, and wind turbines have been destroyed by material failure before, the stresses on these contraptions are real.

The biggest problem with wind energy is its unreliability, the wind doesn't blow continuously, wind turbines only work at certain wind speeds, there are shutdown limits, it is not continuous, and wind speeds vary constantly, and as such the output of wind turbines also varies constantly, and this brings about additional pressure for people that maintain grid stability. Capacity factors of wind turbines fluctuate, mean capacity factor is somewhere around 30%, which means that you may only expect an average output of 30% of the maximum rated amount. That 8 MW turbine actually produces 2.4 MW on average using that calculation. Most of the time, a wind turbine may be spinning, but it is generating very little energy. It requires the correct wind speed range to produce significant output. It is quite common as well to see wind turbines that are not rotating at all. Needless to say its output at that point is zero.

it is worthwhile to note that physics itself limits the performance of wind turbines. A moving fluid (i.e. air) contains energy that is proportional to the **cube** of its velocity. What does this mean for wind power? Imagine an example 1 MW wind turbine, which by design has maximum output in a 40 km/h wind. This turbine would produce 1.00 MW when in a 40 km/h wind. However, in a 10 km/h wind, the energy is $1/(4^3)$ or $1/64^{th}$ of the original power. This means that 1 MW turbine would now be producing approximately 0.01 MW. For this reason among others, wind power output is *very* unreliable. As such, there are long periods of zero or near zero power output, even though the average output is close to 30% of the maximum "nameplate" capacity.

The lifespan of wind turbines is pretty limited due to aforementioned issues like fatigue, maintenance, and risk. The oldest commercial wind farm—Cowley Ridge—in Canada is only 23 years old and will be decommissioned. Rebuilding it is economically not viable and as such it will be scrapped... What happens to the turbine blades? The only thing you can do with them is to disintegrate them, end of life, end of story. What happens to the concrete foundation? It will probably remain where it is, for it is too expensive and to dig them out of the ground, and it costs too much energy. And if they choose to do it anyway, they will probably crush the foundation, grind it and use it to make a parking lot somewhere. And then there's always 680 000 tons of steel to be recovered, a difficult and not yet free exercise.

Wind energy is a promise that will never be fulfilled, simply because the meaningful large capacity units cannot be mass-produced, which is a prerequisite given the sheer volume of turbines required to power a civilization of 9 or 10 billion people that need water, food and electricity.

We move on.

You may consider energy from biomass to be one of the prime evils in energy. This sounds blunt, right? It's true! Consider this harsh truth: Most countries and energy companies now understand that realizing the "green dream" is only possible if you take whatever base-load providing source you can find and paint it green. this means converting coal-fired power plants into biomass-fired power plants. Let's use a more accurate term: Tree-fired power plants.

The strange part about biomass is the assumption that which gets emitted during the scorching and burning of wood will be reabsorbed by forests all around the world, and as such it is considered to be a renewable energy source. This is an incredibly fallacious form of reasoning.

Firstly, it takes a lot longer to absorb the carbon than is initially presumed. Secondly, the rending of natural cover (a net loss of mature trees and plants) actually destroys the ability to absorb said carbon emissions. To add insult to injury one has to consider the amount of fossil fuels used to cut the wood, transport it over the roads and oceans of this world to get it to the power plant. The Drax coal-fired power plant in North Yorkshire, England—rated at

Misnomers and non-solutions

4 gigawatts, one of Europe's largest—has been retrofitted to be able to *eat* US east coast trees as well as coal. The trees that are being *eaten* by these former coal plants come from boreal forests of the US, Canada, Norway, Finland, and Russia.

The misbranding of biomass as a renewable energy source is a convenient cop-out for coal-burning energy companies, an easy way to keep their facilities running and keep spewing carbon emissions, the perpetuation of the combustion economy. Think about it. Rather than getting coal out of the ground and burning it, we've now made our coal-fired power plants slightly less polluting but even more destructive. Think about the sheer volume of coal-fired power plants all over the world that may be viable for conversion into one of these wood-eaters, If we continue on this path we would have precipitated a new age of forest destruction without end.

The conversion from coal to wood at the very least keeps a bit of reliable base-load generation on the side of the renewables, but that's about as good as it gets. Consider this simple fact: According to EuroStat in 2013 the renewable energy mix of Europe looked like this: Geothermal 1.5%, Solar 8.6%, Wind 10,2%, Hydro (dams) 14.2%, Biomass 65.5%. Let that sink in... 65.5% of all "renewable" was biomass and another 14.2% was already maxed out hydro, adding up to a total of 79.7% of dispatchable non-wind and non-solar energy.

It should have become obvious by now that the term renewable should be chucked into the bin and should be deemed dangerous if countering Climate Change is the actual goal. Simply consider that in 2013 40 billion pounds of wood pellets (crushed trees) were burned for bioenergy. Is this supposed to be *renewable*?

I'd suggest we stop burning and especially this foolish regression into the pre-industrialized age when the burning of trees and animal dung were our only means of getting heat.

Science a la carte

This is the *green paradox*, to accept the destruction of natural cover under the guise of producing renewable energy.

A short look at the destruction of pristine lands like the Brazilian and Asian rainforests for the cultivation of palm oil, sugar cane, corn and other crops will teach us that this also is unsustainable. These crops are also used to fuel our combustion economy. For instance, bio-ethanol is widely used in Brazil to propel cars. If we account for the loss of carbon sequestration capacity and the emaciation of the land and the emissions that are tied to these processes we see that they impact the biosphere negatively in a huge way. A paper written by Joseph Fargione, Jason Hill, David Tilman, Stephen Polasky and Peter Hawthorne suggests that the carbon debt from burning these biofuels is 17 to 402 times larger than they offset, which means burning less coal and natural gas. You can find this paper called "*Land Clearing and the Biofuel Carbon Debt*" which is available at sciencemag.org.

Biomass is also used to create hydrocarbons (fuel for cars for instance). The issue with these is that they emaciate the soil, require vast stretches of land that would otherwise be used for growing crops for human consumption, use precious water resources, and in the end the final product needs to be chemically refined (such as by acids and catalysts). It is a terrible and incredibly wasteful and dumb process and is yet another ruse to keep money flowing in a feeble effort to keep animating the combustion economy. And, many people buy into these ideas because they have been branded *renewable*.

Note that the expansion in biomass capacity is far greater and far easier than expanding on wind and solar, which make "renewable growth rates" look good, yet the balance slants towards biomass rather than wind and solar. So if anyone ever tells you that renewables are outpacing nuclear energy, for instance - that's true if you include the increased conversion from coal to biomass and the implementation of new biomass-burning facilities. Also, keep in mind that nuclear expansion at this moment isn't really as fired up as it can and should be, which would be orders of magnitude bigger and far more effective.

You may call biomass the black stain of the renewable revolution and this stain keeps growing as the call for renewable energy gets louder, and law-makers and policy-makers apply (convenient) false logic to this paradoxical and destructive technology.

Suppose you work in a factory, you get paid to maximize production the rate of X, the higher the production rate, the better your pay. The production rate of X depends on the constant influx of energy and raw materials. Suppose there's a sudden stop in wind and your factory exclusively works on wind power? Would this aggravate you?

Suppose you are a medical specialist. You are specialized in performing open heart surgery. You are doing it right now, and suddenly everything goes dark, the life-support fails, the lights are down, and you're stuck in there! Your hospital made the "green choice" and installed PV panels on its roof. Right at the *moment supreme* of this life-saving procedure a heavy storm front moves in and blackens out the sun. The implications are obvious. Someone would claim that the hospital would have auxiliary power generation, naturally. But that's the point isn't it?

Enter the next buzzwords for the renewable movement: Smart-Grid, Storage, Efficiency, and Conservation. In order to make wind and solar work in a theoretical grid-context, we need to offset their intermittency. Intermittency is a complicated term for "when the wind doesn't blow, or the sun doesn't shine" Intermittency is defined by what people from the industry call "Capacity Factor", it's the percentage of mean delivered max performance of one of these technologies over a period of time.

Suppose a wind turbine is rated at 1 MW, and it has a capacity factor of 30% it will deliver 2600 MWh in a year. If it would have a capacity factor of 100%, it would deliver 8766 MWh in a year. Capacity Factor is the factor you need in order to calculate the amount of energy you may expect from one of these technologies. The capacity factors of solar and wind range between 15% and 35%. This is also the reason why I do not focus on capacity but on actual generation capacity. The generation capacity is the end-product after an X amount of time, usually a year. So it's the measure of Wh, KWh, MWh, GWh and TWh. The term XWh will be XWh, no ifs, ands, or buts. To me, it's the best way of creating a level playing field on which I can evaluate each technology. I know others like to do it differently, but by now you should have figured out that I am a dissenter. I do things differently.

In order to offset these limited capacity factors people from the renewable movement envision the following: during hours of plentiful generation, you over-generate and put the excess energy in storage and when the generation stops, the storage kicks in and makes sure that electricity remains available.

The first question we should ask is this: how much capacity would be required to build this store of energy? Is it 10% extra capacity? 20%? A higher number? How do we determine this? Should we consult a meteorologist? In his roadmaps, Mark Z. Jacobson of Stanford University claims that we would require 93 GW of CSP and 4.950 GW solar thermal units to generate the excess energy needed to be stored in order to account for peak load. Our current production rate of solar is 50~75 GW per year. So in order to build this capacity by 2050, we would need to build approximately 150 GW per year for storage alone. That's at least twice the amount we build today. This is not an accurate model of reality... Note that according to Jacobson it would suffice to add 32 thousand GW of solar capacity to the grid (which would require a twelve-fold increase of current solar additions). I wonder how Jacobson is going to bridge the gap from actual production ratio and what it should be. Also, note that Mark Z. Jacobson magically reduces energy projections from somewhere around 240 thousand TWh to 110 thousand TWh. This will be put further into perspective later on. Don't forget, there are technologies of which we can quantify how they are going to increase production.

Second question: how do we store this energy?

Energy storage is a process beset with energy loss due to conversion steps, storage leakage and transmission losses. Possible storage solutions are batteries, pumped storage, flywheel storage and electricity to gas conversion. The pumped storage option is probably the most sensible thing to do (as all of the crazy options for storage go). But the issue with pumped storage is that you can only store a limited amount of energy. Consider flat lands like Finland for instance, how would they make pumped storage work without any meaningful elevation? The alternative is going underground, which increases cost and poses more problems. Also, consider the fact that whatever you do to offset the intermittency of wind and solar, you have to add to the accounting of wind and solar. This only compounds the issue even more in terms of money, materials and production, transportation, installment, maintenance and decommissioning.

There is also the idea that we can create smart grids, which are more intricate and even more complex iterations of existing power grids, including load-balancing features, increased switching capabilities, storage, etc. Some environmentalists think that "requiring a grid" is a non-starter for nuclear energy. But what is a smart grid? Is that a non-grid? It still is a grid... In fact, the grid that is required to account for intermittency and fluctuations is far more complex than the one we currently have.

One of the often mentioned storage solutions is hydrogen (usually through electrolysis). This is going to sound quite impudent, but hydrogen is an even more stupid idea than wind and solar. Want to know why? Here's why: To create hydrogen using electrolysis and use it in some wells-to-wheels chain requires too many conversion steps, which means that your end-efficiency will drop dramatically into the low 20% range. This means that if you put 10 TWh into a hydrogen system, you end up with 2 TWh of functional energy. This is a tremendously wasteful process. This is true all across the board, from propelling vehicles—don't burn it but convert it—to using it to regenerate electricity. That's why I am in almost complete opposition to the hydrogen economy idea. One of the very few useful applications of hydrogen would be in airplanes and rocketry. Also, consider that for instance fuel cells aren't made of ubiquitous elements; these are once again chemically driven units that consist of different layers of certain copper and aluminum alloys and membranes and materials such as platinum.

I'd rather see us have a robust and stable and well-maintained power grid that doesn't have to account for intermittency and frequency fluctuations and handling black/brown-outs. The fact that the intermittent energy farming principles call for an increase in complexity of grid systems should give you pause. We should be doing more with less, not less with more.

Also, consider this simple fact, at this moment, the intermittency of renewable energy is predominantly offset by gas-fired power plants. I've expounded upon these issues earlier in "Highway to Dystopia" but it needs to be reiterated. There is not enough storage to store the existing renewable energy and to offset it, so we burn natural gas. We're not moving in the right direction. Implementing renewable generation policies is inadvertently increasing the need for gas-fired energy production as long as true large-scale storage solutions and/or smart-grid implementations are long past due. I would also like to add that the requirement of accounting for intermittency is a big problem that compounds the issues of wind and solar even further. If the solution to intermittency is added complexity, we're doing something wrong because it will only increase the strain on our natural resources. As such, we may conclude that we're on the wrong track and need to change before it's too late.

There are two technologies left for us to explore: Geothermal and Hydro. Hydropower has wrought enormous damages on nature, it has disrupted the hydrological patterns, it has deprived areas from water while completely submerging other areas. Also, consider that is only as reliable as the amount of water it can capture in its basins, which is no guarantee. Simply take a look at the issues at Lake Baikal, or the Hydro Power installations in Nevada, Arizona and California. If the land is as dry as a cork and bathtub rings start forming in your basins, you know that the amount of water is diminishing and that you cannot keep tapping into the kinetic energy of said water. Therefore, we may conclude that Hydro has served its purpose, and will remain to do so, as long as there is enough replenishment. Building new dams or adapting existing ones may increase capacities, but I don't have high hopes and building more hydro only adds to the environmental destruction.

Geothermal, on the other hand, is a convincingly sturdy and reliable source of energy, albeit somewhat limited in deployability, but I will expound on Geothermal in the next chapter where we will examine some of the basic numbers that pose challenges to our civilization.

In conclusion, I would like to explain my stance on wind and solar from a nuanced perspective. I think they are being used for the wrong purposes. I wouldn't use them to power homes and offices in the western world and this is because we already have enough energy. Our grids are very stable and upgrading our energy production to say- nuclear or geothermal, would be a far smarter move since those do not depend on increasingly more complex grids with storage and load-balancing and whatnot. I would, however, encourage everyone to use them to either charge their BEV's (Battery Electric Vehicles) if they live in countries where energy production is predominantly from hard coal and lignite and natural gas. However, this currently is a choice that has more to do with luxury than necessity; most people still cannot afford the BEV or the panels to drive it.

More importantly, I would use these technologies in rural areas and emerging countries where electricity is scarce or unavailable, and these technologies provide a leg-up into a modern age of prosperity and plenty.

If we want to curb emissions, wind and solar can and have to be part of the solution, but deployed in reasonable and smart ways rather than clinging to them as if they are the one and only answer. They have their uses and we have to implement them accordingly. Note that in the end, these technologies will become obsolete anyway. PV and wind are bridging technologies; don't get too attached to them.

I don't think that wind and solar have to be part of the solution, or at the very least, not much more than we already have built. We may shift these technologies from Germany and other large industrial nations to emerging communities and remote areas as a temporary bridging technology and as such help people progress into modernity, while we push on towards the *Thorium age*.

A great challenge for humanity

Mankind uses somewhere between 550 and 650 quadrillion BTU (British Thermal units) of energy per year, this translates to roughly 175 thousand TWh (Tera = 1 000 000 000 000 Watt Hour). How do we generate all this energy? Mainly by burning coal, gas and oil.

A great volume of burners and combustors exists on Earth: There are about 7000 coal-fired power plants in the world, with another 1000~2000 or so planned; there are about ten thousand gas-fired power plants; about one billion cars roam the Earth, practically all of them running on fossil fuels. Consider the volume of other transportation devices: millions of motorcycles and mopeds, tens of thousands of ships and airplanes. And that's not all; think about the volume of coal- and gas-burners that generate heat for homes, offices, and other facilities. Or consider the sheer volume of people that still have to burn sticks and *dried poo* in order to get some energy to heat their water or cook their food.

I'm going to guesstimate a number, suppose that there are about three billion "burners of anything for anything" on the world, that's our challenge. How can steadily reduce this massive amount of burners / combustors towards zero? Is zero even possible? I don't know, but I think it is feasible albeit in a far-away future. Let us speak necessity: we have to stop burning stuff because we must stop carbon emissions. The debt of our 400 PPM (now 404 PPM - March 2016) transgression will be paid in the future. Consider the tremendous energy storage capacity of the oceans and the atmosphere. The ever-growing concentrations of carbon dioxide will increase the heat that is being captured in our atmosphere and oceans. Debt is being added continuously, and a lot of debt has already been built up in the last decades. The climate systems of the Earth possesses a lot of inertia, meaning that even if we would stop burning fossil fuels today, we will experience the effects of man-made climate change for decades to come, and if positive feedbacks have kicked in—and they probably have—even for hundreds of years to come. These are rising oceans, falling water availability, failing crops, civilizational instability, and worse—a loss of biodiversity by species going extinct.

Let's examine another reason why we need to stop the combustion economy:

The World Health Organization (WHO) estimates that about 7 million people die annually from the combustion economy. This combustion economy entails everything from burning dung and wood to propelling airplanes and even rockets. Of these 7 million roughly half die because they are still using archaic methods to get light at night and cook their dinner. Did you know that in Asia Kerosene lamps are still in use? Or that in large parts of the world people use dried animal feces to cook their meals? Millions of people still live in some sort of semi-bronze-age. And to mitigate the amount of deaths from the combustion economy we have to start lifting people out of poverty and help them progress into more modern ways of life. I have to admit that small-scale renewables can help people from rural communities achieve these goals, but when large cities and communities are concerned [consider for instance Lagos, a city of 16 million] a robust grid sustained by base-load energy operations knows no better alternative.

Energy conservation is one of the magic solutions many proponents for powering the world exclusively on renewables conjure up in discussions about energy.

However, humanity is growing and as such we will need more energy to make sure that everyone has access to potable water and food. Also, the need for carbon sequestration—if we want to save life in the oceans—will raise the bar on power consumption even more. And to cap it all off, if we want to curb population growth we will have to eradicate energy poverty, and this will only happen if we provide massive amounts of energy. The EIA (US Energy Information Administration) has predicted that we would need about 250 000 TWh (rounded up) annually by the late 2040's which means that we will be adding another 80 000 to 100 000 TWh of energy demand over the coming thirty years. There are some scientists—such as Mark Z Jacobson from Stanford University—who think that we can lower this figure to say 110 000 TWh, which I find tremendously optimistic. But even *if* we would be able to lower demand to said figure it still poses a tremendous challenge, and I will show you why.

Is energy demand going up or down? Mark Z Jacobson is confident that if we electrify everything we will lose a third of the demand and as such he posits that we would require somewhere in the neighborhood of 110 000 TWh per year in the 2050's. He then takes it one step further saying that India, for instance, will be able to cut their electricity requirements as well by simply electrifying everything, which is an unsubstantiated assumption. In fact, if you look up Hans Rosling we should be helping the people from India to create even more energy production and increase their prosperity in order to curb their fertility rate i.e. curb the population growth. In this way, we would be alleviating stresses on food production and anthropogenic water consumption.

Also, consider the fact that India is an emerging economy that is set to grow, not shrink. This means that they will be expanding their economy by building more manufacturing capacity and more offices and more schools, and they will be modernizing their agricultural practices, and they will expand on mechanized activities—which at the moment are archaic at best. Through these assertions, we can only arrive at the logical conclusion that energy demand is going to rise. And these assertions are true all across the board, simply look at the other emerging countries in Asia and South America and Africa. These countries can slide in at higher efficiencies, but this doesn't mean that it will happen, and it also doesn't mean that the demand will go down; regardless it will be going up unless we are going to force these people to remain semi-industrialized / non-modernized—which in any case is immoral. Who is going to tell them that they cannot increase their prosperity by adding energy generation?

Getting people out of poverty and out of situations of subsistence requires energy and lots of it!

To understand the fallacious reasoning behind the conservation argument I'm going to show you a very simple counter argument. Cars have become much more efficient since they first dawned, in fact over the past twenty years cars have been becoming more fuel-efficient with a staggering pace. Now we're able to buy hybrids with very high fuel efficiencies, some as high as 3 liters per 100 km in contrast to engines that have efficiencies of 10 liters per 100 km that used to be ordinary. We see oil consumption rise despite these increasing fuel efficiencies.

This would be entirely true if these high-efficiency vehicles would be sold all over the world, but they are predominantly available to Europe and North-America and parts of Asia. However if you look at the graphs, you can see the developing countries and emerging economies overtake the developed countries at a high rate, and it won't take long until they consume more fossil fuels than the traditional "west". If you take a look at the graphs provided by BP- or EIA statisticians you can observe the same trends in every energy sector, demand is rising. Fortunately, I am convinced that despite a growing demand we will find ever more efficient ways of producing and consuming energy. However, this does not enforce the *conservation* argument.

A great challenge for humanity

We return to the challenge at hand: Energy demand is going to rise, and it is expected to end up somewhere in the region of 250 000 TWh of energy consumption per year by the 2050's. Let's have a look at the generation / farming technologies and their respective expected annual yields:

- ➢ Nuclear Energy
  Capacity Factor: 90%
  Yield per 1 GW per year: 7.9 TWh
  Energy for roughly 715 000 homes at 11 000 KWh per year per home.

- ➢ Geothermal
  Capacity Factor: 65%
  Yield per 1 GW per year: 5.7 TWh
  Energy for roughly 518 000 homes at 11 000 KWh per year per home.

- ➢ Coal
  Capacity Factor: 60%
  Yield per 1 GW per year: 5.2 TWh
  Energy for roughly 478 000 homes at 11 000 KWh per year per home.

- ➢ Natural Gas
  Capacity Factor: 45%
  Yield per 1 GW per year: 3.9 TWh
  Energy for roughly 358 000 homes at 11 000 KWh per year per home.

- ➢ Hydro
  Capacity Factor: 40%
  Yield per 1 GW per year: 3.5 TWh
  Energy for roughly 319 000 homes at 11 000 KWh per year per home.

> Wind
>> Capacity Factor: 35%
>> Yield per 1GW per year: 3.1 TWh
>> Energy for roughly 279 000 homes at 11 000 KWh per year per home.

> Solar PV
>> Capacity Factor: 20%
>> Yield per 1GW per year: 1.7 TWh
>> Energy for roughly 159 000 homes at 11 000 KWh per year per home.

These numbers define the scale of things. It is the power density argument. Look at a 1000 MW nuclear reactor that provides energy for 715 000 high energy homes and a 1000 MW PV plant that only provides energy for 159 000 homes, a discrepancy of 556 000 homes. Note that I've used the high-end of all the capacity factors.

A 1000 MW solar PV plant requires four million 250 Watt panels. We currently consume about 21 000 TWh of electricity per year; it would require 12 000 of these 1000 MW PV plants or 49 billion PV panels. We add about 50 GW of PV capacity to the world-grid each year, this equals 200 million 250 watt panels, suppose we could double world production rate for PV panels (which is terribly optimistic) we would require 122 years to get them all built. And this is omitting cumulative upkeep! Each panel will have an operational lifespan of about 20 to 25 years, and after that, they will be replaced since chemical processes will ensure that panel-reliability will decline. This means that from the 20th/25th year and on we need to double PV production each year just to keep the rate of growth such that we can make the 122-year mark for building 21 000 TWh of annual generation capacity for PV alone...

Did you know that each PV panel weighs roughly ten to twenty kilograms? And that the amount of copper per panel alone is about a kilo? We would have to use 500 million metric tons of material alone to get all these panels built, omitting structural materials.

We do not have 122 years, because Climate Change is happening now, and the effects of our emissions will last for decades.

The argument for wind energy is the same albeit with some different metrics. Consider Mark Z. Jacobson's 110 000 TWh; we've now focused on PVifying 21 000 TWh, now we're going to windify the rest. We're left with about 90 000 TWh to decarbonize using wind turbines. The most common ones are the 2,5 MW units. Per 1000 MW—or 400 turbines—we produce about 3.1 TWh of annual electricity. 90 000 TWh / 3.1 TWh (per 1000MW) = 29 000 1000 MW facilities and 29 000 1000MW Facilities x 400 turbines = 11 600 000 Wind Turbines. Currently, we add about 50 GW of wind capacity to the mix and this equals 20 000 wind turbines. By this reckoning, we would be building wind turbines for 550 years in order to get the better part of 90 000 TWh decarbonized. Again omitting cumulative upkeep, replacements and a plethora of carbon intense operations in order to get the materials, refine them and haul these large contraptions into place.

Let's acknowledge that the predictions are not 110 000 TWh but 250 000 TWh, so multiply the projections by two-and-a-half.

What about the amount of materials per MW of wind turbine? You mean all 540 000 Kilograms of them? 540 Tons per MW of wind turbine. Building 11 600 000 wind turbines would require 6.3 billion tons of materials, predominantly concrete and steel and comparatively small amounts of copper yet on a tremendous scale, still a lot. Take into consideration that older designs require more materials per MW and that the numbers presented belong to the newer designs.

As such I will submit to you that wind and solar will not be capable of decarbonizing all energy matters on earth despite what some people might have been advertising to you. The scale simply is too big, and also consider the stupefying amount of materials that are required to erect these energy farming devices and the amount of destruction we will have to wreak on the Earth to get them.

If not wind and solar then what?

Hydro, coal, gas and oil are on their way out, so much is clear. Why not Hydro? It disrupts the hydrological cycle and has wrought a tremendous deal of damage all over the planet. This means that we're left with geothermal energy and nuclear energy. A funny fact about these two technologies is that they both exploit nuclear physics in order to transform and move energy. The Earth has plate tectonics and a hot core and this wouldn't be possible without the decay of certain isotopes of Potassium, Uranium and Thorium inside the Earth. This heat rises towards the surface as it is being radiated into the mantle from the core. If we want to utilize this heat in order to generate electricity or use it for some other process we need to get it, and we do so by running fluids through pipes that have been drilled very deep into the surface of the Earth.

There are some difficulties with Geothermal, though, but these are of a different nature. One of the main issues is finding suitable hotspots, places where you don't have to drill too deep in order to get the heat, this adds to the efficiency of the plant and also increases reliability by mitigating possible pipe failures. If you take a brief look at the geology of the planet and the places where geothermal is being utilized we may conclude that it is best used in areas where there are thermal activities near the surface such as geysers and volcanism. One of these places is Iceland. About 89% of their buildings are being heated using geothermal energy and roughly a quarter of geothermal accounts for the electricity production. Because there are some benefits of using geothermal over other energy sources at places of near-surface thermal activity

A great challenge for humanity

I would almost always advocate for the implementation of geothermal over other technologies in these regions.

In a sense geothermal energy is as constricted as hydro energy is. Without the correct geological circumstances, it simply doesn't make any sense to build it.

We arrive at a point at which we are forced to conclude that each energy source has its own particular merits but is also too limited to make a definite impact on a world scale if mitigating carbon emissions and mining requirements are required. Wind and Solar are good bridging technologies for developing countries, combine them with small scale waste-incinerators—like the one that turns excrements into potable water and electricity—and you can start lifting people out of situations of extreme subsistence. By utilizing these technologies we can introduce them to electricity, electricity to use for cooking and washing the clothes. They will also be able to communicate with the rest of the world through the spread of communication technologies.

Hydro is already maxed out, and we shouldn't be adding much more hydro to the mix. We're left with Geothermal (which is limited) and nuclear energy, the scourge for those who are most frightened of it—irrationality, one of the main themes in this book.

We now know that nuclear energy provides the highest yield per MW of installed capacity. This means that it also requires the least amount of MW's installed in order to satiate world energy demand. We're now going to up the ante and are going to have a look at the 250 000 TWh generated per annum challenge.

Consider this energy mix: Nuclear 60%, Geothermal 20%, Wind 10%, Solar 5%, and Hydro 5%. These numbers are entirely arbitrary just as those from Mark Z. Jacobson's Feasibility studies are, but they will serve a purpose, simply to show you that humanity truly is facing an Apollo 11 moment.

- ➢ Nuclear Energy
  60% of 250 000 TWh = 150 000 TWh
  19 000 GW total capacity required

- ➢ Geothermal Energy
  20% of 250 000 TWh = 50 000 TWh
  8 800 GW total capacity required

- ➢ Wind Energy
  10% of 250 000 TWh = 25 000 TWh
  8 100 GW total capacity required

- ➢ Solar Energy
  5% of 250 000 TWh = 12 500 TWh
  7.400 GW total capacity required

- ➢ Hydro Energy
  5% of 250 000 TWh = 12 500 TWh
  3 600 GW total capacity required

These numbers show the clear correlations and discrepancies between base load / high capacity factor energy technologies and intermittent / low capacity energy technologies. Even though the share of capacity and annual generation between geothermal and wind and solar are inversely correlated to each other, they all share about the same required capacity with a discrepancy of roughly a million MW and yet geothermal alone provides 1/4th more energy than wind and solar combined.

Let's create some more perspective.

We haven't established a true nuclear building baseline yet, and technical proposals are being made that will greatly enhance building and deployment rates. Suppose we can produce about two hundred 300MW units per year, by this reckoning, it would take us 316 years to get 250 000 TWh constructed. We would be adding 60

A great challenge for humanity

GW to the mix each year. Let's add another 50 GW in larger units to the mix - this means fifty 1 GW reactors adding up to 110 GW of annual nuclear additions per year or 867 TWh, again we're going to do the division game: it would still take us 173 years to get it done. And this is throwing big punches!!! Does it start to sink in? People advertising 100% Wind, Water, Solar are truly out of touch with reality. Even I find it challenging to make due on nuclear within a reasonable timeframe, and this is by far the most potent energy source we have. We literally need an energy miracle!

Suppose we want to achieve full decarbonization of energy by 2050 (34 years from now) using said energy mix.

- ➤ Nuclear Energy
  150 000 TWh / 24 years = 6250 TWh per year
  Equal to 792 GW per year
  1 320 0.3 GW units per year
  396 1 GW units per year

- ➤ Geothermal Energy
  50 000 TWh / 24 years = 2083 TWh per year
  Equal to 366 GW per year
  1.219 0.3 GW units per year

- ➤ Wind Energy
  25 000 TWh / 24 years = 1042 TWh per year
  Equal to 132 GW per year
  135 000 2,5 MW units per year

- ➤ Solar Energy
  12 500 TWh / 24 years = 521 TWh per year
  Equal to 66 GW per year
  1.19 billion 250 Watt units per year

Let's get back with our feet on the ground and figure out what we can do in the foreseeable future. Suppose that we can at least decarbonize electricity by 2050. The challenge will be somewhere between the 25 000 and 35 000 TWh mark. Let's assume that our civilization by then consumes 35 000 TWh of electricity per year, and we are going to try doing it using the same energy mix (nuclear 60%, Geothermal 20%, Wind 10%, Solar 5% and Maxed out Hydro 5%).

- ➢ Nuclear Energy
  21 000 TWh / 24 years = 875 TWh per year
  Equal to 110 GW per year
  185 0.3 GW units per year
  55 1 GW units per year

- ➢ Geothermal Energy
  7 000 TWh / 24 years = 292 TWh per year
  Equal to 51 GW per year
  171 0.3 GW units per year

- ➢ Wind Energy
  3 500 TWh / 24 years = 146 TWh per year
  Equal to 48 GW per year
  19 000 2,5 MW units per year

- ➢ Solar Energy
  1 750 TWh / 24 years = 73 TWh per year
  Equal to 41 GW per year
  264 million 250 Watt units per year

Where the first figures really made me glum, these make me confident that at least electricity generation can be decarbonized using this mix within a reasonable amount of time. In fact, wind and solar are less strained in this scenario than these technologies currently are, which would actually give us some room! We can use these technologies to bridge gaps here and there, helping transportation become carbon neutral, or helping emerging people to move towards energy prosperity.

A great challenge for humanity

The 55 figure for 1+ GW plants might seem daunting, but in the year 1975 we were already starting to build 43 units and before that we had four years in which we started building more than 35 units. So the 55 unit figure from a technological standpoint has to be feasible. The bottleneck would be foundry capacity. At this moment very few places in the world are capable of building these single piece, thick walled reactor cores, these facilities are only available in a handful of places in Japan, China, France and Russia. Luckily heavy foundry capacity is set to grow by additions of capacity in China, South Korea, Russia, Japan, the Czech Republic and England. Without these foundries, we would not be able to add enough nuclear capacity to the grid in order to achieve our goal of full decarbonization within a meaningful timeframe unless there are designs that can be built without the need of a specialized very big foundry. Another consideration in the mind of the innovators is the scale of the components, they want them to be as small and as modular as possible so that the *ease of transportation* and *replaceability* of these components increases and with it the operational lifespan of a facility. It will also provide a platform for ongoing material science discoveries and continuous technological advancement of individual components to increase reliability and lifetime and efficiency.

How reasonable are these expectations?

China is planning to grow from 27 GW of nuclear capacity in 2016 to 150 GW in 2030. This is a growth rate of 9 GW (9 000 MW) per year, which is 16% of the required worldwide 55 1 GW units per year, and 8% of the 110 GW requirement. Luckily there are dozens of other countries that are also setting their eyes on nuclear expansion.

France added 58 reactors over a 20-year time span, which means about three each year.

Science a la carte

In 1973 we started building 42 nuclear reactors all across the world, suppose these were all 1 GW units, we would be just shy of 13 GW away from our 55GW 1 GW unit quotum.

About 45 emerging countries are considering nuclear energy according to the World Nuclear Association. The limitations on nuclear expansion (depending on the technology used) are not determined by cost and/or materials, but the amount of nations that are willing and *brave* enough to do it. Using the word brave is for those who think that nuclear energy is scary, I would rather use the words *rational* and *calculating* and *foreseeing*.

Thorcon is a business that wants to replicate the success of the MSRE (the Molten Salt Reactor experiment), and they estimate that they can build 100 1 GW units per manufacturing facility per year. Terrestrial Energy envisions a similar process albeit for smaller reactors, I don't think it is impossible at all to reach a meaningful growth curve in nuclear additions. We also have FliBe Energy and Transatomic power that have slightly more complex yet also modular MSR designs, and then there are also contenders like the full burner design of Terrapower, Prism, the AP1000, APR1400, EPR, and tested and proven PBR designs. Since we now have dozens of companies working on their own designs, we will probably see an increase in manufacturing capabilities and some of these designs will be modular and can be mass produced. I think it is justified to be optimistic about a nuclear future given the clear superiority in energy density and material requirements. Simply think about the revolution that was wrought on the world of production by Henry Ford in 1908 by introducing the assembly line. The modularity of new reactor designs is precisely in tune with this philosophy and will bring about the same revolution albeit roughly 110 to 120 years later. Consider this question: Has it been done before in the nuclear industry? The answer is clearly no, and that is the reason why we should be optimistic, especially because there are companies that are planning for it. Once this gets going and the price for nuclear energy drops below the cost of coal, the world will switch around and eventually the *Thorium age* will begin. It will be the logical next step in utilizing nuclear fuel to energize civilization.

Note that James Hansen, the well-known climate scientists from Columbia University has posited a required implementation rate of 115 nuclear power plants per year in order to mitigate carbon emissions. With my 110 GW per year, I don't think I'm that far off. We have to start soon, though, because it might seem daunting, but it is absolutely necessary if man-made climate change is to be defeated and if we want to lift our collective civilization into a new era of optimism and stability and meaning.

There is one last fact to consider: a reactor vessel ranges from 300 metric tons to about 750 metric tons, and this means that for each wind turbine you build, you have enough steel to build a small reactor core which has about 100 times the capacity in megawatts. This is the true measure of success, doing more with less. Note that there's a large discrepancy here, building 250 reactor cores is far less energy intense than building equivalent wind capacity, say 60 000 units. It is these differences that matter the most since they will drive the demand for raw materials down considerably. The material requirements for building nuclear power plants is a drop in the bucket compared to building wind and solar in order to reach the same end-result over a period of 100 years. If we would take the championed 100% WWS path, we would be engaging in a far more wasteful process than these people dare to admit. I expect them to disagree with me strongly. However, I will let you and those who understand the nature of evidence to be the judge of that.

Besides, if we would engage on this path of nuclear development and innovation, we shall arrive at a point where we will be able to increase deployment speed. Also, we would be taking the path of energy prosperity, and this prosperity would also greatly enhance the search for off-world resources, and a transition into space. We might start with prospecting asteroids or the moon or Mars, but we will never know if we would start to regress due to energy poverty, which is a logical outcome of the 100% WWS philosophy. Also, consider this to be my ultimate "decoupling" argument. As Mike Shellenberger, for instance, says "*better by far to decouple civilization from nature than to live in harmony with it.*" Nature is the single most potent life-force on the planet; it is the system that keeps our planet habitable. It is time we learn to leave it alone as good as possible so that it may flourish and keep life on Earth possible. I do think that some harmony is possible, but nature is not there to serve us, we may behold and preserve it.

If we could manage to decarbonize electricity by the 2050's we would already be making significant headway. And by introducing these deployment figures, we would be ramping up the learning and deployment curves tremendously. If we would ask statisticians and researchers to do the math based on physics and available resources, they would concur, of that, I am quite certain, for I have done the math myself...

Let's reiterate the importance of decarbonization so it will become clear. To help the biosphere recuperate we need to stop burning fossil fuels and biofuels because they contribute to mass denudation of precious living lands and because they interrupt the carbon cycle and introduce too much greenhouse gases into the atmosphere. Also, these processes are particularly harmful to life in general, and this is exemplified by the annual reports provided by the World Health Organization. Reconsider for instance the mind-boggling fact that each year about 7 million people die from the effects of the combustion economy.

And as if all of this isn't harrowing enough we are facing an era of massive civil instability and strife, of ever increasing wars and refugee crises and eventually—if we don't pay attention—mass extinction through acidifying oceans and increasing droughts and famine. One of the main causes for all of this is the way we consume energy. We have to stop *burning stuff* and learn how to mass implement solutions that both alleviate stress on nature and provide us with the energy required to feed an ever expanding human civilization and to keep progressing into a more optimistic future. I submit to you that the 100% WWS approach that some advocate doesn't work and that we have to include new nuclear developments and innovation in order to set things straight. This is a challenge of great proportions, but one we can meet!

One final metaphor: Would you send a fighter into the ring against a champion, with his hands tied behind his back? That's what the sages in favor of 100%WWS are trying to do, start the Bennie Hill theme, please...

The efficiency argument

The correct implementation of efficiency is going to lead us into a future of plenty and prosperity. The implementation of the most efficient technologies will ensure a faster transition because it serves to lower the threshold at which we may progress; it lowers the strain on resources and electricity demand. We're going to examine a couple of simple situations to explain why efficiency is so important. I promise you that this will be the most boring of all the chapters in this book, but a necessary moment of dullness because it will drive home a powerful point.

Let's do some back-of-the-envelope calculations—Google Enrico Fermi...

I shall be superfluous because we need some extra insight so we may learn from it.

Magazines and websites dedicated to reviewing and providing information about cars often cite carbon emissions per driven unit (mile or kilometer) and most of these state zero emissions for BEV's, but is this actually the case? Yes and no... What can we learn?

Suppose we have a Tesla Model S 90 KWh and a ford fiesta 1 liter Ecoboost with an efficiency of 1 liter per 23 kilometers, and they both drive 20 000 kilometers in a year. Each car is driven as economically as possible. You can gun either vehicle and its fuel-efficiency will drop dramatically. We don't want that, so we're aiming for max range.

The efficiency argument

Fiesta 1 Liter Ecoboost

Fuel Tank = 42 Liters | max range = 42 x 23 = 966 km
Emissions = 99 Grams CO_2/km = 95 634 Grams/Tank
20 000km = 20.7 fill-ups x 42 liters = 869.6 liters consumed
20 000km = 20 000 x 99 Grams CO_2/km = 1 980 000 Grams CO_2 emitted

End-result after driving 20 000 km in a Ford Fiesta on high fuel economy: 1980 kilograms of CO_2 emitted.

Tesla Model S 90 KWh

Capacity = 90 KWh and max range = ~500 KM
Total energy used = 20 000/500 X 90 = 3600 KWh
EIA figure for lignite is 2.17 pounds of CO_2 per KWh
3.600 x 2.17 = 7812 pounds
1 pound = 0.453592KG
7812 x 0.453592 = 3543 KG CO_2

End-result after driving 20 000 km in a Tesla Model S 90KWh: 3543 kilograms of CO_2 emitted.

The Tesla emits:

3543 - 1980 = 1563KG CO_2 more when driven on 100% Lignite
3380 - 1980 = 1400KG CO_2 more when driven on 100% Coal
1980 - 1922 = 58KG CO_2 less when driven on 100% Natural Gas
1980 - 0 = 1980KG CO_2 less when driven on Nuclear or any other non-emitting energy source.

There's a point to be made against electric cars, as long as coal is used the most to generate electricity even the BEV is worse than the gasoline car, however, I think that the BEV will win in the long run, and it is evident isn't it? Let's think it through one more step, let's add some context.

I live down south in the Netherlands, right between Germany and Belgium. The energy mix of the Netherlands is dominated by coal and Gas (83%), the energy mix in Belgium is dominated by nuclear energy (53.7%) and the energy mix in Germany is once again dominated by coal (47%) and gas (6.4%) and biomass (10%)—believe it or not, in 2014 Nuclear still generated 17,2% of all of Germany's electricity, more than wind, more than solar, and more than wind and solar combined... If I lived in Belgium, I would consider buying a BEV without a doubt. If I lived in Germany, I would not consider buying a BEV unless I knew that Germany would re-invest in nuclear energy, and the same can be said for the Netherlands. There's one way to counter this issue and that is by buying solar panels and charging the car during the day, which for me is a possibility as I do my writing at home, but others don't have that luxury. To me, this is the most important consideration for buying a BEV: "Where does the electricity come from?" And as long as the answer is predominantly fossil fuels (or biomass) the BEV isn't as good as one would think. Enter one of the reasons to keep pushing for low-carbon energy sources and particularly new and modern nuclear power designs.

What about average cars? I own a 2004 Volkswagen Bora 1.6 16V. It is rated at 168 Grams CO_2/Kilometer, and it achieves a fuel efficiency of 14.3 kilometers per liter. This means that it will emit 3360 Kilograms of CO_2 per year per 20 000 Kilometers. This is only marginally better than a BEV run on lignite, so from this perspective it is easier to knock off some KG's off your annual vehicular emissions, drive a 250 Grams CO_2 per kilometer car and you're better off with the BEV... This is the juxtaposition either drive your BEV on nuclear energy and be the king of the road in terms of lowering emissions, and if you drive your BEV on coal you're average at best, but there isn't really that much choice is there, moving is your best bet, or try to convince your government that nuclear is the way forward, which I am going to do. It's the slow route, though, yet absolutely worth it.

What else can there be said about the BEV and the combustion engine?

Gasoline contains roughly 10 KWh/Liter (edit: 8.9 KWh/Liter based on EPA figures). The total energetic value used in the Fiesta after 20 000 km = 869,6 Liter x 8.9 KWh/Liter = 7739 KWh. So basically the Tesla is more than twice as efficient with its 3600 KWh.

What if we are going to synthesize the fuels for these new gasoline cars? We're going to have to generate 7739 KWh per car per 20 000 km's rather than 3600 KWh per car per 20 000 km. And it is uncertain whether synthesizing this fuel is 100% efficient. Suppose synthesizing fuel for gasoline cars is 50% efficient, this means that we need 15 478 KWh per car per 20 000 km rather than 3600 KWh.

The 3600 KWh figure is not entirely accurate either; we need to generate the electricity, transport it and convert it. Transportation is about 10% loss at the most, AC/DC conversion another 10% which means that the wells-to-wheels efficiency from the point of transmission is 81%, and this means that we need 3600 / 81 x 100 = 4444 KWh per Tesla 90 KWh per 20 000 km—consider that the engine efficiency is about 90% as well but is already figured in because of the range / KWh figure. Also, note that a figure of 4444 KWh actually makes the BEV perform worse than has been pictured before. But I will leave the numbers up there because I am keen to see who pays attention.

How many reactors would we need to decarbonize all personal transportation based on cars? The South-Korean APR1400 is expected to deliver roughly 11TWh per year, that's 11 000 000 000 KWh

11 000 000 000 KWh / 4444 KWh = 2 500 000 90KWh tesla's driving 20 000KM per year.

If we want to decarbonize all personal vehicles (approx 1 billion) by converting them to BEV's we would need: 1 000 000 000 / 2 500 000 = 400 APR1400 reactors, that's not a lot. For that reason I submit to you that the electric car, in the end, will be the correct choice when paired with nuclear energy, they are in fact the perfect low-carbon match.

This is the final statement that should convince you that nuclear-powered BEV's are the way into the future. Nuclear energy costs between 50$ and 150$ per MWh. As we progress, this figure will drop, in fact, Terrestrial Energy has proposed a reactor design that will generate electricity at a cost between 40$ and 50$ per MWh. Let's take the 50$/MWh figure since this has been realized in France, for instance.

400 x 11 = 4 400 TWh / 4 400 000 000 MWh per year.

4 400 000 000 MWh x 50$ = 220 000 000 000 (220 Billion) $ per year in operating costs for Terrestrial's IMSR at 50$/MWh propelling 1 000 000 000 Tesla Model S's 20 000km in a year.

1 000 000 000 Fiesta Ecoboost cars use 1 000 000 000 x 869.6 liters = 869 600 000 000 liters of gasoline annually. How much does it cost to create 1 liter of gasoline?

There are multiple explanations, and I'll run with one them. "*De Rekenkamer*" a Dutch television program invited several academics who deduced that 1 Liter of gasoline should cost about 65 euro cents. There are a lot of factors that determine the cost, oil price, labor, refinement costs, transportation costs, etc. Let's run with this number.

869 600 000 000 x 0.65 Euro = 565 240 000 000 Euros.

The current exchange rate (April 20th, 2016) is 1.13 dollars for 1 euro.

565 240 000 000 x 1.13 = 638 721 200 000 Dollars.

The costs will end up between 600 billion dollars and 700 billion dollars for a high gasoline price. If the price of oil rises, so does the cost per liter of gasoline produced. This is a world-scale day-to-day business in which oil prices determine who buys which oil from who and in this way oil is exchanged a lot.

The BEV wins by great margins in terms of efficiency, cost and environmental impact when driven on nuclear energy.

These are oversimplifications; the reality is far more complex. Consider a plethora of different car designs with different efficiencies. We now looked at it from the perspective of a small car with a small engine running on relative high fuel-efficiency. The overall cost for fueling 1 billion cars all around the world (cost of gasoline, diesel, and gas) is probably much higher than the 640 billion dollars per year figure.

That's all fine and well, but there are a lot of people who tell me that hydrogen is the way forward, what about hydrogen?

There are currently two viable ways to create hydrogen: first, by steam reformation of natural gas, effectively making 80% of all hydrogen produced and used a fossil fuel. Second, we produce hydrogen by electrolysis of water. Electrolysis is the much-touted solution to create the green future; it is the pure electrical means to get hydrogen gas. Consider the electrolysis chain and wonder if it would be better than the BEV chain:

Transmission (90%), AC/DC conversion (90%), electrolysis (50%), compaction / liquefaction (90%), fuel-cell (50%), electrical engine (90%). I am omitting pumping and purifying the water before it can be used in electrolysis, but the point is clear.

The total efficiency of the fuel cell electrical vehicle (FCEV) wells-to-wheels chain is somewhere between 15 and 30%, let's take the higher figure, I'm feeling generous as I write this. Suppose you want to drive 20 000KM with the FCEV. We know that the electrical engine will drive 20 000KM on 3600KWh of in-battery-stored-energy and so it is 3600 / 30 x 100 = 12 000 KWh required to drive a hydrogen car the same distance as a BEV will go on 4444 KWh (wells-to-wheels) which is a discrepancy of roughly 7500KWh per vehicle per 20 000km.

How many APR1400's would we need to fuel the FCEV future? 11 000 000 000 KWh / 12 000 KWh = 917 000 FCEV's driving 20 000KM per year. If we want to decarbonize all personal vehicles (approx 1 billion) on 1 000 000 000 / 917 000 = 1090 APR1400 reactors contrasted by 400 for the BEV future—nearly 700 more.

Each step in the hydrogen chain brings about loss, and this is the basis for the efficiency argument—look up the laws of thermodynamics. Besides the hydrogen economy very much looks like the oil economy, after refinement, all this gas has to be carted around, because we all know that for the sake of profit it is far more economical to produce hydrogen in large quantities in giant hydrogen plants, rather than at the point of consumption. Either way, point-of-consumption-generation would mean transmission losses whereas centralized generation would mean transportation losses, both are opposite sides of the same coin really, the coin of inefficiency. Remember, if you want conservation all across the board, you have to stay true to those principles; you cannot have them a la carte...

Numbskull politics

A constant in human existence has been the use of subterfuge and fabricated truths for one's own benefit. It began quite early during the stone-age, when there were people who didn't understand the natural world and had superstitious tendencies. These tendencies led to the creation of shamans and witch doctors and eventually morphed into what we know today as religion and pseudo-science and—in some cases—politics. The perpetual hoodwinking and selling of snake-oil and fooling eludes us even today. In fact, it is doubtful that it will ever end. But we have a chance to hold these people accountable and we should.

Consider the thirst for wealth and power present in many individuals of the human species. The things we do to gain wealth and power are often abhorrent. We kill and maim, steal, and lie. Mostly the acquisition of wealth knows no boundaries in terms of human and natural cost.

"Is it about nationalism or environmentalism?"—Bill Gates - 2011 WIRED Business Conference

An excellent question! Let's briefly and superficially look at the things certain nations and parties do that are either paradoxical or even outright dangerous.

COP21—a two-degree non-solution

Let me put it bluntly: The COP21 accord is hopelessly inadequate to address the carbon-debt wrought by 200 years of combusting and burning hydrocarbons. If this is the measure of our resolve, we're in deep trouble. The collective scientific community has been raising the alarm, especially in terms of Ocean Change / Acidification and the possible Arctic Methane Apocalypse. The doom of the world as we know it will be decided by our willingness to leave the value of money and ideology behind and base our decisions and our future course on empirical principles, evidence-led thinking, rationality, and reason.

If you are sensitive to the profane I would suggest you to avert your eyes and skip this paragraph but in my view the *"let's limit the earth's warming to two degrees"* agreement is political numbskull fuckwittery of the highest order. This is taking a gigantic gamble with the future of generations to come—your children and mine—how many generations? We may expect that civilization as we know it will end if we do not put a halt to the disastrous change we've precipitated on the oceans, the arctic, and the biosphere. Well before it destroys whatever we hold dear on this planet, it will become very messy and very bloody.

Are those who struck this accord aware that pushing the limit up another two degrees might take us beyond irreversible tipping points? This should be one of the prime concerns!

The debt has already risen to such an extent that we have to start remediation immediately. Latent effects are already in the pipeline, and our governments have capitulated to greed and decided that we may put even more emissions into our atmosphere and oceans, but at what cost?

Besides all that, who is going to police this agreement? There's no true judicial obligation if a country wouldn't make it, or by some weird twist of faith would be forced into the opposite direction, no-one is going to stop them and the agreement will be in tatters. One may also note that there are no technological viewpoints in the COP21 agreement.

Numbskull politics

Are there any provisions in this COP21 agreement? Have they agreed on certain conditions at which there needs to be done more? When is anyone going to pump the brakes? It is truly alarming to note that there's no real focus on the state of the Earth's climate and that the main focus remains on economic development, rather than true and immediate mitigation. I have to add this little caveat here, economic development for emerging economies is important, and it will serve to lift the populace of these countries into a more healthy and knowledgeable and prosperous state.

"We really have an emergency because of the inertia of the system"—James Hansen at COP21

What does Hansen mean with inertia? Have you ever seen a flywheel? A flywheel stores rotational energy in its mass and this energy can be used to start an engine, if you put a lot of energy in this flywheel, you can't simply stop it by stopping to feed it energy, it will keep spinning until you've taken out a lot of the rotational energy that it has stored, which either means letting it spin for a long while or applying brake pressure. It's the same with the climate, even if we would stop feeding it with energy, it will remain spinning for a long time to come because it has stored masses of energy and it keeps getting more energy than it is releasing. Unless we restore a relative equilibrium, the climate will become more unstable over time.

Consider the effects we are seeing right now. Climate Change has already catalyzed massive movements of people by exacerbating discontent and the instability of Northern Africa and the Middle East for instance. The seemingly unending refugee streams flowing into Europe stem from water shortages and crop failures which in term destabilized already trembling societies. Also, consider that these countries have seen shortages for years. Agricultural Syria was all but decimated in the years leading up to the great civil war. People from Bangladesh and Eritrea had been suffering from water shortages for years, and now they've had enough. These people will be seeking for lives of plenty and stability and safety elsewhere, and so they become migrants or refugees.

As the Glaciers in the Himalayas are shrinking and the monsoon periods are a changing we might expect increased water shortages in the Asian Subcontinent for many years to come. In addition to all these issues, we may note that India does not have a well-developed water system which will add to the pressure of the problems that already loom on the horizon. Becoming a destabilized country of a Billion inhabitants will have severe repercussions for the rest of the world.

Even though the COP21 agreement might be heralded by many as historic and a good thing, I will deem it nonsensical and totally missing the point. Despite all the harrowing possibilities our consumption may have precipitated, world leaders have chosen to prolong and potentially even worsen them. As such I will call the COP21 agreement a disaster. Economics and geopolitics have trumped reality and rationality.

Sweden—willfully sabotaging its own carbon-free energy

The energy mix in Sweden is largely clean-air and almost totally non-carbon emitting. Their nuclear reactors produce about 50% of their energy; roughly 30% of their energy comes from hydro; 8% of their energy comes from wind. This leaves us roughly 12% of thermal / unknown energy sources. Given the fact that the graphs for the unknown sources represent a flat line we can deduce that those are not PV panels because generation graphs corresponding with solar look jagged.

We may conclude that Sweden is one of the proudest countries in Europe, right? I mean they have a tremendously robust, safe and plentiful energy generation system that supplies base load power not just for themselves but also for their neighbors if need be, and a generation system that doesn't severely damage the biosphere, nor pollutes the atmosphere with hazardous elements.

There's an interesting article in the Bulletin of the Atomic Scientists: "*How to decarbonize? Look at Sweden*"

Excerpted from its abstract: "*To light the way forward, we need to examine success stories where nations have greatly reduced their carbon dioxide emissions while simultaneously maintaining vigorous growth in their standard of living; a prime example is Sweden. Through a combination of sensible government infrastructure policies and free-market incentives, Sweden has managed to successfully decarbonize, cutting its per capita emissions by a factor of 3 since the 1970s, while doubling its per capita income and providing a wide range of social benefits.*"

They did this by building nine nuclear reactors.

What is their plan? What has made me turn my attention briefly towards this Scandinavian country? They are raising taxes on nuclear energy! But to what end? If you take a look at the link shared below, you can see that the entire Scandinavian and Baltic energy network is interconnected and that Norway, Sweden, and Latvia are mostly exporting energy to Denmark, Finland, Estonia, and Lithuania.

http://www.svk.se/en/national-grid/the-control-room/

Today, for instance, the figures are these:
Sweden - **exports 2864MW**
Denmark - imports 266MW
Norway - exports 219MW
Finland - imports 1348MW
Estonia - imports 309MW
Latvia - exports 207MW
Lithuania - imports 911MW

What can we learn from this? First, these are net numbers; certain areas of Sweden for instance import energy from Norway, whereas Norway in other regions imports energy from Sweden. Sweden exports energy to Denmark, Germany and Poland, and this goes on and on and on all around the Baltic Sea. So how reasonable is it to look at electricity in this region as merely a national issue? What would happen if Sweden gets rid of its reliable base load providing nuclear power plants and supplants these with non-dispatchable renewables such as wind turbines? The answer is clear, it won't be able to help its neighbors as effective anymore. Subsequently, their neighbors will be forced to eventually look towards gas-fired power plants to generate the electricity that they need.

With the taxation of nuclear energy, the Swedes may well drive the profitability of nuclear energy over the cliff and then what? What is their plan B? Sweden is tiny in terms of energy consumption and the case may be made that wind additions and pumped storage will provide enough energy to offset the loss of nuclear energy, which is something the Swedish green party envisions.

Even though this is entirely feasible this will give the wrong impression, it will add to the worldwide fervor of non-sequiturs surrounding renewable energy: "Look at Scotland!" "Look at Denmark!" "Look at Norway!" "Look at Sweden!" "Look at Germany..."

First, Germany has not decarbonized, in fact, it has added burning capacity; Second, countries with relatively small populations and low energy requirements are easy to decarbonize using wind and solar, however, ask yourself this question: what is their backup? Where does the energy come from when there's no sun and no wind? Denmark and Scotland get their base-load from their neighbors. Third, Norway is quite unique has fjords...

Numbskull politics

Germany—A foolish and failed quest

The Germans, I love them, I live about 2 Kilometers from the German border. I've visited quite a few football matches in Germany. It's a neat country, their houses are nicely arranged, you mostly see pristine German cars driving around, and their bratwurst and beer are excellent! Having said that, I also think they are quite foolish.

In 2002 the Germans decided that they wanted to overhaul their electricity system. Coal and nuclear were supposed to go out, and wind and solar were supposed to be the new pillars upon which the strongly industrialized German nation would have to be supported.

Up until now the only thing they've phased out is a part of their nuclear capacity; subsequently, they've added biomass and coal and gas and massive amounts of *renewable*. However, this renewable increase did not amount to much in the larger scheme of things, generation wise there are only very marginal additions, amounting to not more than 15 percent of their total electricity generation and this is still being trumped by nuclear regardless of having been phased out partially. Even if the Germans would phase out all nuclear energy, chances are that they will have to increase their import from Sweden, Poland, the Czech Republic, Switzerland, and France... and keep burning coal. How much wiser would it have been to keep their nuclear reactors running and slowly but surely phase out their burning capabilities?

On a world-scale, we might have seen a net increase of carbon-neutral generation capacity but the total sum carbon-neutral actual generation has gone down disproportionately and that is due to the closure of nuclear reactors and the inverse relationship with wind and solar.

The funny thing is that the Germans are probably becoming aware that they cannot turn off the rest of their nuclear reactors because we may note that solar additions, for instance, have been slowing down and seem to stall soon. Wind is still growing, but this growth is marginal and not as forthcoming as to guarantee or even warrant a complete transition away from nuclear.

What is the goal of the Energiewende? Is it to curb emissions? Is it to get rid of nuclear? Is it to get rid of coal? They have been able to shut down some nuclear, so if that's one of their goals it has been partially achieved. If their goal is to shut down coal? They have failed miserably. Besides, they already rely heavily on importing the energy production of their neighbors, particularly France and Poland. Do the Germans expect to become more reliant on them in the future? Hadn't they shut down their reactors, they wouldn't have needed to import as much electricity as they have done.

The true goals the Germans aim for is a reduction of 80% of their electricity emissions and 80% renewables by 2050. The first goal is excellent, yet consider this, how much better would it be to shut down lignite- and hard coal-fired power plants and supplant them with nuclear and renewables? Even though I would say that implementing nuclear would be the smartest thing, I do acknowledge that it might be safer politically not to do so, but this shouldn't really matter... And I generally hate this idea...

An expertly written blog post on this issue is available on actinideage.com called "*The Lightbulb Moment*". It is absolutely true if curbing emissions were the objective shutting down nuclear power plants is the dumbest thing you can do. Yet consider that 43% of all the Germans think that phasing out nuclear energy is a good thing and a reason to support the Energiewende and so they have set out to shut down a perfectly fine and excellently operating electricity generation technology and have thus chosen coal and biomass and gas over nuclear energy. I wonder if they ever realized that this choice has probably killed thousands of people in Germany and neighboring countries. Coal has been proven to kill hundreds of thousands of people each year, consider for instance James Conca's article on Forbes *"How deadly is your Kilowatt?"* and wonder how misguided Germany's choice was.

The final nail in the coffin comes from an article published Pushker Kharecha and James Hansen on the NASA website called *"Coal and gas are far more harmful than nuclear power."*

Biomass is one of the "renewables" that is often weaseled in there in order to make renewable figures look better than they really are. In Germany this figure fluctuates between 7,5 and 10%.

In the end, the economics of the Energiewende will force the Germans to abandon this quest. In fact, there are reports that suggest that Germany is going to stop subsidizing wind energy and as such we will see the most potent force of the Energiewende end by 2019. By that time the Germans will have invested roughly 1.1 trillion US Dollars and have achieved a very marginal success, in fact, if you look at their mitigation targets we may not even call this a success at all, au contraire, it may be considered a failure. For 1.1 Trillion they have installed 43 GW's of wind capacity which by now satiates about 9.7% of their total electricity demand. According to Fraunhofer and the BMWi (Federal Ministry of Economic Affairs and Energy - Germany), wind capacity has grown from 12GW in 2002 to 43GW in 2015, an annual growth rate of 2.4 GW (7.4TWh), hardly exciting. Solar has grown slightly more in the same period, but it has an even worse capacity factor than wind, so even if we concentrate on solar we still see a growth rate of 3 GW (5.3TWh) per year.

The volume of natural gas capacity has also increased, which is unsurprising because that it is the preferred technology to offset the intermittency of Wind and Solar. The Germans have added 8GW of gas-fired energy generation over 13 years, 0.6GW (2.2 TWh) per year. Brown coal has remained stable, and hard coal has slightly declined in capacity, it has lost a mere 2 GW's over 13 years, which we could deem hardly noticeable if curbing emissions is the objective. Energy demand in Germany has been growing since 2002 and to satiate this demand they have grown from ~117 GW in 2002 to 185 GW in 2015, an increase of 68 GW.

The final stick to beat Germany with is its investment in Biomass, it has grown from 1.5GW in 2002 to 8.5GW (43TWh) in 2015. It has offset the loss of capacity in hard-coal. Nuclear has declined from 24 GW (189TWh) to 11 GW (87TWh), which is the regression and the reason why the capacity has risen as it has, they had to make up for the loss of 100 TWh of annual generation capacity coming from nuclear...

They gained about 250 TWh of generation capacity, lost 100 TWh to nuclear, so their net increase was approximately 150 TWh. The Germans could have achieved the same thing by building 17 AP1000 reactors. Instead, they have built an equivalent to 12 400 2,5 MW wind turbines and 156 000 000 250 Watt PV panels each with operational life spans of about 20 years, contrasted to the 60 years that you may expect an AP1000 or APR1400 to operate.

Let's contrast these figures with Terrestrials IMSR projection of 50 dollars per Megawatt hour produced. The German Energiewende officially started in 2002; it will end in 2019 which means that they have spent 1.1 trillion over a period of seventeen years or 58 billion per year. For the Germans, this is not a huge figure because their economy is the biggest in Europe and they have money to spend. however, if you take a look at what can be done with that money and contrast that with what has been done, we may only conclude that this money has been wasted. As Ben Heard would say, this was a very expensive experiment that has shown us what not to do with our *precious* money. Any idea what 58 Billion US dollars can buy you? The Germans could have decarbonized their entire electricity grid by now and have electricity to spare, to share with their neighbors—if only they took the IMSR way at 50$/MWh. To be fair, this option is not yet available, even today albeit not far away.

I don't think it to be unreasonable to suspect that the Germans will choose such a path in the end. Before the 2020's are over the Germans will have built new reactors, and I am betting that one of them will be a molten salt reactor. But before that happens, many euros will be spilled, and more coal and trees will be burned because no one will switch off the lights and stop the economy.

France—an inexplicable regression

France is known for its baguettes, great assortment of cheeses, the Eiffel Tower, the city of love, and June 6th, 1944 [D-Day]. One might think that it is a laid-back, easy going country—which is true—but it's also a powerhouse of technological progress and innovation. As if CERN and ITER aren't impressive enough, France has been the frontrunner in a showcase to prove how reliable and robust nuclear power can be. In a matter of decades, they transformed their electricity generation capacity. Archaic coal-fired power plants were replaced with standardized and institutionalized nuclear reactors which by now have been able to supply roughly three quarters of the annual electricity consumption in France. It has also helped the French keep their emissions per capita amongst the lowest in the European Union.

Isn't it baffling that they—like the Germans—think that they can easily manage without their nuclear reactors and they, under the leadership of a left-leaning government, are planning to transition to something else: renewables. I would rather encourage them to prolong their leadership in nuclear research as they have shown in the seventies and eighties, and continue to show with the ITER and CERN initiatives. Why would they choose to regress? It's ideological; there can be no other explanation for it. Let's take a look into what leads the French to go astray?

Consider this publication on the website of Scientific American:

"*France Loses Enthusiasm for Nuclear Power*

Nuclear's share of electricity will drop from 75 percent to 50 percent by 2025 due to loss of know-how and requirements for more renewable sources."

What does "requirements for more renewable sources" mean? Does it mean that it is a policy decision to implement more renewable? What is the rationale behind this decision? To sum things up: they have lost expertise, Areva hasn't been profitable, and President Hollande made a campaign promise to edge France away from nuclear energy and its inevitable "waste" issue. It still seems very strange, even though the French produce far more energy than they actually consume they choose to turn it down.

Consider this strange idea, they want to pave their roads with "*solar frikking roadways*" (YouTube). It sounds like a wonderful idea: "*Let's remove all asphalt from the roads and replace them with photovoltaic panels so that it can capture electricity while the sun shines and no cars are there to block out the sun. We have millions of miles of road anyway.*" It's the same argument all over again, how much surface area is required to capture a meaningful amount of energy? How much materials are required per unit or per surface area? What do we need to do to get these materials? What will it take to maintain this network? How much replacing do we need to commit to? What is the reliability of the road-surface? How safe is it? What about failures?

Also, consider this simple fact, if we are keen on recycling we have to acknowledge that we will be trading in one of the most recycled materials on the planet... Asphalt. Even though Asphalt is a byproduct of the petroleum industry, it can be re-used almost indefinitely, which means that we probably have enough of it to keep perfecting our infrastructure for decades.

Where the French have decreed that all new roofs have to be covered with plants and/or solar energy capturing devices. I would say: scrap the idea of solar panels and go for solar heating modules which can be constructed without the added hazardous chemical byproducts that photovoltaics have and also consider that the materials required for these units are far more ubiquitous. Solar heating modules serve to reduce the use of gas and electricity in terms of heating your homes, the dish- and bath water. The net gain of solar heating modules would be far greater than of Photo Voltaic units, simply because they remediate the energy requirements and have far greater latency and storage capacity and basically serves to achieve more efficiency rather than try to provide small and meaningless portions of intermediate energy.

The question remains, what kind of government will be required to make the correct choices. I think it is proven that a far right or a far left leaning government isn't the answer. In fact, politics is driven by fear and emotion rather than empiricism and as such we need to find the most rational people and try to get them into office. Rather than having a congress of lawyers, populists, and economists we could do better with far more engineers and scientists. People that can help steer legislation onto a more balanced course that is free from ideological unsubstantiated numbskull decisions. Rather than scrapping all nuclear facilities, France could easily remain a frontrunner by embracing new technologies like MSRs for instance. They are a technologically advanced country that can make these impressive innovations possible and keep showing the world how to run nuclear energy safely and reliably and cheap as they have done for so long.

Science a la carte

Let's put these facts together and consider the consequences. The two biggest economies in Europe are dependent on one and other in terms of energy. In fact, Germany has to import energy in order to keep the grid stable. Also consider the fact that Germany has grown its "green energy portion" by means of burning more biomass—which inadvertently kills forests and releases yet more carbon dioxide and other pollutants. their share of wind and solar is meager, simply consider the fact that Great Britain—which doesn't advertise a to be a "champion for green" actually has more wind capacity than the Germans. Also, consider that the growth potential of renewable energy is limited because of their rare ancillary material and chemical requirements, and to offset the loss of their nuclear generation capacity they have built new [lignite] coal plants and gas burners. What will happen as soon as France's grid gets turned into a renewable grid where base load has gone? Which of the two nations is going to supply the energy required to run their economies? The surrounding countries only have very limited generation capacities and are also trying to participate in the ideological *green* push for renewable energy... Also, consider this simple fact: France doesn't import any energy, it only exports it, and its biggest customer is Italy. Suppose France's generation capacity goes down and with it their export capabilities, Italy will be forced to build more capacity, and this will most probably be gas, don't for a minute think that it will be renewables exclusively as we have *just* learnt that their intermittency needs to be offset by fossil-fuel burning... if we are going to trade French nuclear reactors for Italian gas-fired power plants, we again are set to regress.

Many environmentalist groups (& the politicians they influence) are falling for the sideways logic of Jacobsen and Oreskes and their ilk. Good intentions do pave the road after all, but where will this road lead? Through the dark and haunted forest? It sure seems so. Oreskes has chosen the path of antagonism by calling "us" "renewable deniers" and conflating this stance with "climate denialism." Let's be honest, having written the stuff that I've written so far, have I given you the impression that I deny the causal factors that lead to exacerbated climate occurrences i.e. anthropogenic climate change? Oreskes needs to come back to earth and adhere to the principles she advocates as an academic.

The Republican Party—A gathering of science denying and cherry picking loons

We have arrived at one of the most ridiculous political movements of this decade: the Republican Party of the United States. What these people do is beyond me. I can't fathom how they choose to defy science, yes defy, not deny. Consider for instance Jim "the snowball" Inhofe, or any of the presidential candidates of the 2016 election, particularly Ted "cooking bacon with guns" Cruz. [edit: Ted Cruz has yielded to Donald Trump and is no longer a candidate — May 2016]

Most of the time I have a very dim view on politics, but when the US is concerned and in particular the republican party—since George Bush Senior left the office it has become a caricature of itself—I almost start to cry, sometimes from laughter, other times from near despair. It's a collection of Looney Tune figures molded from real people, performing a compendium of ludicrous Monty Python sketches. Each Senator is crazier than the rest. This whole grand old party is a collection of anti-social, theocratic, pro-life, creationist, science denying, gun-loving, money grubbing, oil loving, war mongering, curmudgeons and cretins and jesters. I love the literary exercise, and I could go on forever, but we have to learn something from it.

If you consider the Republicans of yore, the likes of Richard Nixon, Ronald Reagan, and even George Bush Senior, you see people that were [slightly] in touch with reality. Nixon, for instance, acknowledged the harm anthropogenic processes have had on the biosphere. Reagan, on the other hand, was a disaster on environmental issues and would have perfectly fitted in with the republican clown-corps of today. Nixon on the other hand:

"The Congress, the Administration and the public all share a profound commitment to the rescue of our natural environment, and the preservation of the Earth as a place both habitable by and hospitable to man."—Richard Nixon

This is something I doubt any modern day Republican would say, because of the implication of seemingly acknowledging the truth behind man-made climate change i.e. leaving the safety of the herd...

"'Environment' is not an abstract concern, or simply a matter of aesthetics or personal taste—although it can and should involve these as well. Man is shaped to a great extend by his surroundings. Our physical nature, our mental health, our culture and institutions, our opportunities for challenge and fulfillment, our very survival—all of these are directly related to and affected by the environment in which we live. They depend upon the continued healthy functioning of the natural systems of the Earth."—Richard Nixon

Richard Nixon's legacy on environmental issues is good, he signed off on legislation and new agencies that have become instrumental in protecting the environment:

- The National Environmental Policy Act
- The Environmental Protection Agency (the EPA)
- The National and Atmospheric Administration (NOAA)
- The Clean Air Act
- The Clean Water Act
- The Endangered Species Act

Where it is true that Nixon was in a sense a person with a disastrous legacy, there are some redeeming elements to his presidency. However, one of the most disastrous decisions he made was the elimination of the MSRe project, which could have had a great and positive impact on our energy consumption and would have modernized nuclear energy many decades ago, and would have mitigated a lot of the damage we've done ever since. Is he redeemed? Yes and no... The pivotal technological innovation that could make a real difference got trashed by a personal decision made by Richard Nixon.

"*It kind of reminds... I could use the Third Reich, the Big Lie. You say something over and over and over and over again, and people will believe it, and that's their strategy... A hot summer has nothing to do with global warming. Let's keep in mind it was just three weeks ago that people were saying, "Wait a minute; it is unusually cool*.""—Republican Senator James Inhofe

Apt use of World-War II rhetoric, Nazi-scare—Godwin's rule of Nazi analogies? I guess that Inhofe is a particular fan of hyperbole.

"*Well, actually the Genesis 8:22 that I use in there is that "as long as the earth remains there will be seed time and harvest, cold and heat, winter and summer, day and night," my point is, God's still up there. The arrogance of people to think that we, human beings, would be able to change what He is doing in the climate is to me outrageous.*"—Republican Senator James Inhofe

I rest my case... God did it, God does it. We humans cannot mess with the plan of the almighty. Right... I only have to quote these characters! Is this really a serious argument?

"*The authorizing committees are free to set their agency budgets, and that includes **NASA**.*" - Republican Senator Ted Cruz

The thing that worried me the most about Ted Cruz was that he was trying to defund NASA's earth-science programs as chairman of the United States Senate Commerce Subcommittee on Space, Science and Competitiveness. This includes satellites that observe the earth and its mechanisms and help us understand our planet more and anticipate what is going to happen. The observations made by NASA are unequalled and without them we would go blind. But now... Cruz is running for President of the United States, which means he could get executive powers if enough idiots vote for this blithering cretin and that could even be more dangerous because he could do a great host of things that could endanger the biosphere even more, it almost keeps me awake at night... Luckily, it didn't happen, yet... Cruz may always run again, there may always be someone with the same convictions. We have to safeguard ourselves against these cretins, how can we do it? It's a question that pains my mind occasionally.

"To the revolutionary understanding that all men and all women are created equal. That our rights do not come from the Democratic Party or the Republican Party or even from the Tea Party. Our rights come from our creator"—Republican Senator Ted Cruz

Should this man be given the nuclear launch codes? Is he a rapture Christian? Is he rational enough to be President? I think his sanity should be tested first before allowing him into the Oval Office. I do think it prudent to incur checks that prevent lunatics from getting to sit in the mightiest chair in the world, for too much is at stake.

"If you look at satellite data for the last 18 years, there's been zero recorded warming, The satellite says it ain't happening."— Republican Senator Ted Cruz

Which satellites? What data? Who has done the analysis? Have you spent money on scientists in order to get this analyzed? What about all the people working at NOAA and NASA and the IPCC and the countless of research institutes all over the world? Are you telling me that they have doctored all the evidence? The same data that has probably come from the same satellites you were referring to? You can say *"satellites do not support the warming 'theory'"* but when you do so, you're expected to present evidence that supports that claim. Did you provide evidence? Or is it merely a mantra you keep repeating in your silly little echo-chamber, Ted?

"It is altogether worrisome when you have scientists treating matters — denouncing those pointing to the actual facts and data as deniers. And indeed, I would point out that was the exact same conduct the Flat Earth people demonstrated toward Galileo. And the global warming alarmists in their treatment of those looking to the facts and evidence often behave like modern day Flat Earth proponents."—Republican Senator Ted Cruz

Trying to claim moral superiority by conflating your own dissenting view with the facts that Galileo presented isn't going to make your case any more true or false. This is a meaningless meme you're using. There is no parallel between Ted Cruz and Galileo. In fact, if you analyze it, Ted Cruz is far more like those who were scorning Galileo for his support of the Copernican system, later to be known as Heliocentrism or the notion that the planets of our solar system orbit the Sun. He also conflates criticism of those that deny or defy science with persecution which is ludicrous. If your notions are proven to be erroneous, and you're in office and don't alter these ideas which you present as untouchable dogma, you should be held accountable. I don't feel comfortable with any raving lunatic in office passing legislation on matters that may determine our capabilities that help us understand our world better, or alleviate our vexing influence upon it. If you are not being held accountable, or at least addressed as being someone whose views are out of touch with reality, sadly enough we're going to regress, and there are plenty of people that have been convinced by the likes of you [Cruz] that man-made climate change isn't happening. Again man-made climate change is substantiated by masses of evidence, far more evidence than your little brain is capable of comprehending. Is this ineptitude of yours the origin of the ludicrous conspiracy theories? Let's be reasonable, do you [Ted Cruz] really think that tens of thousands of academics and teachers and science communicators have doctored the evidence to hoodwink good and hardworking conservative constituents in order to steal their money?

We have yet to reach the top of the lunatic GOP king-of-the-hill movement. Let's see who Republicans think is a credible running mate in the presidential race...

"At those times on the campaign trail when sometimes it's easy to get a little bit discouraged, when, you know, when you happen to turn on the news when your campaign staffers will let you turn on the news ... Usually they're like "Oh my gosh, don't watch. You're going to, you know, you're going to get depressed.""

"The chant is "drill, baby, drill." That's what we hear across this country in our rallies because people are hungry for those domestic sources of energy to be tapped into. They know that even in my own energy-producing state we have billions of barrels of oil and hundreds of trillions of cubic feet of clean, green natural gas."

"I don't think the public gives a flying flip if somebody knows who, today, is a specific leader of a specific region or a religion or anything, because that leader will change, of course, when the next president comes into power, just based on the volatility of politics in these other areas."

"Trump's candidacy has exposed, not just that tragic — that ramifications of the betrayal of a transformation of our country"

"ultimately what the bailout does, is help those who are concerned about the healthcare reform that is needed to help shore up our economy uhm helping uhm, oh! It's got to be all about job creation to shoring up our economy in in putting it back on the right track so health care reform and reducing taxes and reining in spending

has got a company tax reductions and tax relief for Americans and... trade we have it we've got to see trade as opportunity not as a competitive, scary thing but one in five jobs being created a in the trade sector today we we we've got to look at that is more opportunity all those things under the umbrella job creation this bailout is a part of that."

One page of sheer lunacy is enough, especially after having seen the mind-boggling stuff Cruz and Inhofe have shared with the world. Behold the queen of garbled speak: Sarah Palin. If Senator John McCain would have won the presidency back in 2008 she would have become Vice-President, I kid you not! She would have been one misfired heartbeat away from being President...

Where practically all nations on earth are in general agreement that man-made climate change is a serious issue, the Grand Old Party—also known as the Republican Party—denies it and their reasoning is flawed and unsubstantiated and if they retain power in the United States of America, they might very well exacerbate the problems, rather than alleviate them. They want Americans to keep consuming fossil fuels because it delivers, it is a means to get money and therefore leverage. The beauty of it is that Republicans like to play the victim card. It's the persecution-complex, they all claim to be the Winston in their own 1984-narrative, and we are the thought-police... It's this kind of irrational and conspiratorial mindset that perpetuates harmful and idiotic practices in the world's most powerful and influential country, and this is inherently dangerous. Republicans love to use the word *hoax*, and it actually gains traction with their constituents, and this deepens the chasm between the Republicans and the Democrats, which leads to all sorts of antagonistic yet irrational counter movements from either side.

The Republican Party is blessed with numbskull politicians that dare to cross the border into the realm of the ludicrously stupid.

Science a la carte

Whatever one of them does, there's always one standing by to up the ante. If there are still sane people left in this political movement, they have become irrelevant; they're only there to create leverage, nothing more. What is even worse is that the narrative does trickle down from the senate floor to the common man. The amount of people that have swallowed the republican narrative hook, line, and sinker is immense, and these cognitively impaired people make sure that this political facade can go on forever. Let's hope that sanity returns to the realm of US politics. I would love for both the Democrats as the Republicans to return to their senses and focus on science and reason to be their guides whilst steering the United States into the stable future that the country deserves. Will it be a path of regression or a path of progress? Which is not merely a matter of Democrats versus Republicans; For this to happen requires a change in the frame of mind, a turn towards rationality and evidence-led thinking. The inspiring greatness of the space age is what I am looking for!

Democrats also have blemished themselves by killing off the IFR; by opposing Vermont Yankee; and by stifling political consensus regarding Yucca Mountain. The opposition of nuclear energy and the overly optimistic view on renewables, well... Let's just say that there's something wrong on either side of the aisle.

It doesn't matter to whom you kneel, whether it is some dear leader or a non-existent deity; it also doesn't matter who levies the taxes or imposes his or her will upon you; climate change doesn't give a damn about all these issues. Climate Change knows no ideology nor doctrine nor dogma or allegiance, it just is; and we have made it become a dangerous—and in the end quite possibly a deadly—monster.

As such I submit to you that it is childish nonsense to politicize Climate Change. Even though I am a utopian, someone who wants everyone to live lives of plenty and safety and stability, I don't care about political ideology, I am not a joiner-upper anymore, and neither should you. We should demand empiricism, science and reason in our day to day dealings, rather than feelings and popularity ratings and political pageants...

Ken Caldeira has an interesting suggestion on this matter "*It would be interesting to do a study of climate sensitivity estimates and correlate them with political viewpoints. Marty Hoffert has hypothesized that there is a correlation with people on the right estimating a lower climate sensitivity than people on the left. This would be an interesting study in the sociology of science.*"

There's still much we can learn and I am confident that no physicist or engineer would run a country or an energy system based on ideology and confirmatory and cognitive biases. This existential threat of ours does not take place at the low level of communities or countries, we're facing an existential threat on a world-scale, and therefore, ideology and borders and numbskull politics do not matter.

Boundaries are not evident when you behold the Earth from outer space. A maxim often enunciated by the wise Carl Sagan, one of my personal heroes of all time.

The pale blue dot is all we have...

The worth-nothing of ideology

Many people are cognitively closed to reality. For some reason, they seem to be unable to acknowledge the situation we're in; the fact that we live in a state emergency; a state in which we cannot disregard anything; and a state that forces us to accept all non-carbon-emitting technologies. We have to push on into the realm of continuous innovation and improvement, upward towards an era in which we—the humans—can live on this planet in harmony with and yet separate from nature.

By omissions of vital realities and the failure to re-examine their reasoning, these cherry-picking individuals, and groups, in fact, disqualify themselves from having any say in the future of our planet and civilization, we simply cannot have our science a la carte. Let's have a look at those that show a clear incompetence to address the emergency we face.

Many of these individuals accept science A (the complete Climate Change compendium) but not science B (for instance the possibilities for nuclear energy to progress). It's cherry picking really. Especially when they do not address the shortcomings of the technologies they themselves advocate. Who are these people? Why do they make me question the validity of their arguments?

350.org—what is it, are you pro- or anti-nuclear?

"There has never been a better moment in history to break free from fossil fuels." & *"In 2016, the fossil fuel industry faces an existential crisis."* & *"There they will engage in **civil disobedience** to stop the digging in one of Europe's biggest open-pit lignite mine"* & *"There will be coaches travelling from all over Europe, and training beforehand."*

These sentences have all been mentioned in an electronic newsletter I received on March 3rd, 2016 from Tim Ratcliffe of the 350 organization. Firstly I agree that the time for fossil fuels to go has come, but I recognize the gargantuan role they play in our day-to-day energy matters and therefore some questions come to my mind: *"What is your plan B?"* and secondly *"What technologies are we going to deploy to supplant these lignite coal plants?"* and *"Have you thought about the efficiencies and limitations of these technologies?"* and *"Are you also going to show the same civil disobedience where nuclear power is concerned?"* I question whether they truly know what is required to supplant coal-fired energy generation with anything else.

Also, note that James Hansen is listed as one of their chief advisors, a person who actively endorses nuclear innovation. Is this also a-la-carte activism from 350.org's part? If 350.org does endorse nuclear energy, why not mention it?

The answer came a few weeks later in an e-mail from the hand of the same person (Ratcliffe):

"Friend,

*Nothing like Break Free has ever been attempted before: mass, coordinated actions in every corner of the globe to call an end to fossil fuels **and usher in a 100% renewable future**"*

76

Science a la carte

Which brings us to an important topic of this book: Cognitive Dissonance and the breeding thereof.

Within a communal species like Homo Sapiens cooperation is geared towards mutual benefit and survival, we humans tend to choose groups based on trust and well-being. We tend to immerse ourselves in, and and start following the pervasive mantras, dogmas and doctrines of these groups. It is basic *monkey behavior*. I am confident however that we are slowly and gradually but surely freeing ourselves from these archaic modes. But we're not there yet, not by a long-shot.

Consider these outspoken anti-nuclear activist groups and individuals: Bill Mckibben, The Precourt Institute—Funding research aimed at postponing the implementation of true solutions; Greenpeace—Destroying Nazca Lines and showing falsified anti-nuclear lunacy; Sierra Club—an ideological shift from opposing dams to opposing nuclear energy; Friends of the Earth, Helen Caldicott, Naomi Klein, Amory Lovins, friends of the earth, Mark Z. Jacobson, The ecologist, Dana Durnford, Jim Green, Vandana shiva, the Union of concerned scientists, and finally US presidential candidate Senator Bernie Sanders. This is a tentative list. Let's hope that one, or preferably, more individuals change their minds in the course of time.

Admittedly there is also some misalignment between several of these groups and individuals, but there's a general focus on protecting nature, fighting climate change and also fighting or questioning the value of nuclear energy. In this chapter, we are going to home in on the hypocrisy of some of the aforementioned people.

Let's find out if these people are really serious about combating climate change. I will mine some quotes (how disingenuous), and I will rebut them.

The worth-nothing of ideology

— Bill Mckibben, writer and "climate change activist"

Bill also writes e-mails on behalf of 350.org. The following quote, however, has been excerpted from another article.

*"I don't think **thorium** or **cold fusion** or anything like it is the future of power; I'd wager all things nuclear are mostly relics of the past, in no small part because they **cost like sin**. But the point I was trying to make is that the new fact in the world is the remarkably rapid fall in the price of renewable energy. That solar panels cost so much less than they did just a few years ago strikes me as a destabilizing factor for anyone's world view."*

On the cold-fusion part, I tend to agree, I do not think that it is ever going to be proven to work, my skepticism comes from the fact that it is ridiculously hard to make atoms collide with each other. We can make them collide in machines like the Wendelstein 7-x (Wendy), which is a proof-of-concept for *hot* fusion. The energy required to get this process going is many orders of magnitude bigger than what cold fusion is supposed to do. Note that Wendy is already running tests and proving that fusion is possible, the learning curve is steepening. Why is it so hard to fuse atoms? The cores of atoms consist of protons and neutrons. Protons repel each other, they "don't like" to be next to one and other. Once two protons are in collision course with each other, they will swerve and miss. So we have to create immense heat and immense pressure to ionize them and make them collide and subsequently stick together. This is incredibly hard to do and cold fusion simply does not facilitate in the energy levels required to fuse atoms.

On the other hand, Mckibben is mistaken when he thinks nuclear a thing of the past. When we consider the past of nuclear energy and acknowledge the fact that there has been little true innovation, we may only conclude that there are huge steps to be taken in the nuclear industry. Consider the fact that our current fleet of nuclear power plants mainly consists of generation I and generation II reactors and that rigid and expanding regulatory pathways have caused prices to soar and innovation to stagnate. The stagnation in the nuclear industry has been imposed upon it by bureaucracy. The fact that we have not yet any commercial molten salt reactors is not because the MSRe was a failure; it is because the US chose not to continue with the technology and basically bury it for decades, which means that it is imposed bureaucratically rather than technically or economically. Are there perhaps other reasons to shut down the MSRe? It certainly seems so, for one it was situated in the wrong area, Nixon wanted nuclear development to continue in California rather than Tennessee. Secondly, the MSRe didn't have a strong role to play in weapons production because it is generally very proliferation-resistant. The MSRe got scrapped because of geopolitics and politics. Note that the MSR probably provides the easiest pathway to utilizing Thorium as a nuclear fuel. The aforementioned 'probably' meaning that there might be other technologies that could be faster or easier to achieve but have not been known to me yet. I do know that India is working on Thorium reactors since they have vast reserves of the element and they need enormous amounts of energy to modernize their country.

And then there's the futile "*solar only costs a penny*" argument. You can have all the pennies of the world but if you cannot get the materials required to *do it* you're stuck, aren't you? These are the basic facts why solar and wind are doomed to fail on a grand scale: We simply cannot get enough rare materials required to build enough solar panels; Production ratios are limited for a reason... Also, we have to minimize mining activities; this means choosing the most effective technology that has the highest power density and operates reliably.

The basic question that occupies my mind is "*why is Bill Mckibben so adamant in opposing nuclear energy?*" If the answer is either money, Fukushima, bombs, Radiation, and/or Chernobyl? We'll simply have to draw the conclusion that he is either misinformed or dishonest. A person writing about nuclear energy must be able to answer those questions correctly, and I will.

Finally, take note that Bill thinks that the price-tag is relevant, why should we worry about price while the stability and health of the biosphere are concerned? Even though I am convinced that the economics are in favor of nuclear energy, it shouldn't matter. To say that we cannot save civilization and mitigate its effects on the planet because it costs too much is a stupid and ignorant position. Money is a nebulous and temporary ill of society, I am convinced that money will disappear in the future, probably far ahead, besides any bank can press a button and conjure up the money you need to get anything done, it is called credit...

"Sorry we wanted to save humanity but we didn't have the money to do it."—Me, just now...

— Helen Caldicott: Pediatrician extraordinaire

Consider Helen to be the queen of black and white argumentation, false causes and slippery slopes.

"No dose of radiation is safe. Each dose received by the body is cumulative and adds to the risk of developing malignancy or genetic disease."

Here we have a black and white argument, there is no safe dose, therefore, any dose "might" lead to developing malignancy or genetic disease. My argument, based on the reports I've read is this: Radiation is deadly, and radiation is healthy, and everything in between, just like aspirin can be deadly. We need to put these issues into perspective, but Helen chooses the path of intellectual dishonesty by presenting the black and white argument.

Helen has had a considerable anti-nuclear run, she's one of the most vocal people to oppose nuclear. As a person with a medical background, she oftentimes focuses on the medical risks of radiation. Consider the previous quotation. Let's ask a couple of simple questions: Is *each dose received* measured over a day, a month, a year or a lifetime? Is there any moment at which the "risk-counter" is being reset? How big is the risk from natural background radiation? What about the *Linear No Safe Threshold hypothesis* that is about to be falsified? What about Hormesis for which there is actual evidence? Should we avoid nuclear medicine and scanning methods? How is it possible that nuclear medicine cures several ailments if no dose is safe?

Consider for instance that radioactive Iodine or Technetium is used to diagnose and cure Thyroid conditions. It's the same stuff that may cause Thyroid cancer, but it can also be used as a targeted therapy. Now there's a lot context that needs to be added. Simply consider this, as long as Helen doesn't present any credible cohort-type epidemiological research as evidence, her own statements can be dismissed through basic logic and without evidence. Her arguments are therefore moot in my opinion.

Helen weasels in the words "adds to the risk" which can be translated into may, could, with a great degree of uncertainty, etc. etc. Also, consider this question: how big is the risk in the first place? As a doctor, she should know that practically all cells of a human body get replaced over an X amount of time. Also, acknowledge that background (natural) radiation absolutely dwarfs any—accidental—emissions from nuclear power plants or other ubiquitous nuclear isotopes in man-made contraptions.

Helen, please provide the epidemiological evidence for all the cancers and diseases that have been caused by working in the nuclear industry, being a flight attendant or an astronaut. Since these people are subject to higher doses of radiation than the average human being, they should show elevated cancer levels.

The worth-nothing of ideology

Also, consider this; radiobiology is a complete area of expertise in the medical world. All sorts of doctors need x-rays (Google Wilhelm Röntgen) and organ function scans to diagnose ailments. None of these doctors say before you get one of these scans *"we're going to subject you to, or, inject you with ionizing radiation, you might contract cancer from it"*. In fact, if you ask them they say *"the dose is so small... no need to worry about anything"*. Would these people really say this if they had the same convictions as Helen has? Of course not! Why not? Because they know a lot about radiation and its effects on the human body.

What about reducing the amount of radioactivity on Earth? Sounds strange right? But it's true! While we mine Uranium and transform portions of it into energy, we actually decrease the radioactive mass on the planet. And as such nuclear energy can actually be seen as a decrease in worldly radioactivity, rather than an increase. Crazy! I know...

Also, consider this claim: Nuclear power has prevented Hundreds of thousands, perhaps even millions of deaths from air pollution through offsetting the burning of coal and lignite. Simply ask the Germans what they did when they decided to end their dependence on nuclear energy, right, they started burning more coal and trees to get their energy.

I feel quite confident in stating that nuclear energy has saved far more lives than any other energy source, simply by offsetting fossil-fired energy generation, by providing base load energy, and by supplying isotopes for the medical world. One caveat though, the industrial revolution has significantly raised life expectancies all over the world. So at first the massive burning of fossil fuels might be seen as a boon for humanity, but these redeeming qualities are now a thing of the past since we have to acknowledge the reality that is called anthropogenic climate change and the deadly effects it brings to bear on a large and worldly scale.

Why is Helen travelling around the world using airplanes? "*I've been travelling the world for forty years*" (I don't suppose by boat... And she live[s][d] in Australia and the US...) By now she has subjected herself to so much additional background radiation that she should have cancer all around her body. Or is she somehow magically shielded or resistant to it? There can be no other explanation! What's the paradox here? Do I sense a small hint of hypocrisy? End of cynicism.

"*Nuclear power __contributes__ to global warming.*"

Let me guess, there are emissions from mining uranium, processing and refining it, building reactors, keeping them running and then decommissioning them, right? Guess what? The same can be said for any other power source on Earth, in fact if you look at it from an XWh per year per tonnage materials viewpoint, and consider the building costs in materials, lifecycle and end-of-life-cycle it actually turns out that nuclear is by far the least material intensive energy source that there is and one may therefore conclude that its carbon footprint is also least of all. I guess this goes over Helen's head. The "contributes to global warming" argument is incredibly flimsy and when put into context and is contrasted against the means of farming energy—biofuels, solar and wind—dissipates into thin air. In the coming chapters, I will take us deeper into these matters as we are going to compare the different energy sources based on the figures of the IPCC and the EIA. I'll give you a small hint—if it hadn't dawned already—Helen is full of it...

"*The real costs of nuclear power are prohibitive (and taxpayers pick up most of them).*"

The costs of trying to run the world on renewables exclusively—like these demagogues wish would happen—will be famine and societal instability and death. In terms of costs, it is perfectly justified to pay upfront since it effectively mitigates these severe issues. I would deem it highly immoral not to pay a large amount upfront if it could effectively mitigate thousands, maybe even millions of deaths annually. The cost of death and severe illnesses should be factored in with the examination whether energy technology to implement.

Consider the simple fact that in 2015 the Carbon Emissions didn't stagnate due to the feeble addition of say ~100GW of wind and solar, no, it stagnated because energy demand in China was lower than expected because of a drop in production—economic pressure—and thus less coal was burned. Energy demand in China is dominated by the production of goods, a slight economic downturn leading to less export and therefore less production means that they can "shut down" a coal-fired power plant or two, and we have lowered the carbon emissions, right? Note that China has roughly 950 GW of coal capacity and this figure is still growing. Consider that the addition of wind and solar are worldwide. Also, note that 100GW of wind and solar equals about 50 GW of coal-fired energy generation capacity and 33 GW worth of nuclear energy generation capacity.

There's about 2000 GW of coal-fired capacity installed worldwide and according to the World Coal Organization, it is expected to grow to about 3000 GW by the 2040's. And that's the real cost we have to factor in. Are we really going to oppose nuclear and try to go all-in renewable? Do these people actually read energy reports? Or did they attempt to answer some basic physics questions and mathematical issues? Organizations like the World Resources Institute paint even more grim pictures...

Worldcoal dot com expects there to be 2600 GW of coal-fired capacity by 2025. This would constitute a net increase of 600GW over nine years and 66GW per year. So instead of defeating coal, we're losing the battle, so much is clear, and renewables are not making it to first base, so it's time for nuclear energy to walk up to the plate and start batting. This is the only true rebuttal to the "it costs much argument". Cost is determined by many factors, most of which have nothing to do with the actual building costs, but it doesn't determine potential and feasibility—this is the fundamental mistake many people opposed to nuclear energy make.

Science a la carte

If you calculate the amount of deaths and diseases that are prevented by running a nuclear reactor rather than a lignite-fired power plant of the same magnitude you have to acknowledge that the medical cost savings alone serve to justify paying for base load providing and sturdy and reliable nuclear power, upfront.

"There's not enough uranium in the world to sustain long-term nuclear power."

Is this based on the assumption that we can only run nuclear on Uranium 235 (less than 1% of the Uranium Isotopes) with the reactors that we have today? First, according to Transatomic Power, there's enough energy left over in spent fuel to power our entire civilization for another seven decades. Given the fact that we'll never build enough reactors to utilize all of it, based on the amount of spent fuel we may deduce that there's enough spent fuel alone to last for say at least 100 to 200 years. Secondly, If we then extend the fuel cycle to include Uranium 238 (99%) and Thorium—of which we have vast reserves already—we may expect to extend nuclear capabilities by many many thousands of years. By then we should have either cracked the code of producing energy through the principles of nuclear fusion and/or have the capabilities to get Uranium and Thorium from extra-terrestrial sources. Also, consider that by then we will have far more advanced technological capabilities, different materials and a better scientific understanding of the world, which is the same argument the greens pose when saying *"there's still a lot to be gained"*.

The *"there is not enough Uranium"* argument is a counsel of despair. It is the futile attempt to negate the innovation argument; it is a paradox that exposes *green hypocrisy*.

Also, consider that PV technology cannot be recycled completely, which negates the circular economy principle that people would like to apply to it. Want to know why PV technology cannot be recycled completely? Irreversible chemical changes. Besides doping PV wafers is a nasty business, of which many waste streams end up in dystopian places like Lake Baotou in China—In much larger quantities than the nuclear industry has caused during the entire span of its existence.

"Potential for a catastrophic accident or terrorist attack far outweighs any benefits."

Please explain to me how these terrorists are going to attack a nuclear reactor, and what they are going to achieve by it? Also, consider that Generation IV designs—which are just around the corner—are practically full-proof against any of these attacks. Since I love fiction very much, I could imagine many things terrorists could try to do with a reactor? Fire a projectile at it? You could hit the containment dome but not the reactor itself, good luck with that, there's another one or two layers of steel and concrete and water before you even get to the reactor core. You can't steal any material from these facilities in a couple of minutes; getting to the fissile material requires you to either remove it from the reactor or from the spent fuel pool. Neither of which is something you can do within an hour. Then what? Suppose you can get some of this stuff, what can you do? Build a nuclear bomb? Hardly possible without a highly sophisticated chemical facility and without any gamma emissions, which are easily traceable.

Aforementioned Generation IV designs cannot be subjected to any form of sabotage or materials theft while in operation. These operate at such high temperatures, yet so low pressures that you simply cannot get anything out of it. And even detonating one of these facilities wouldn't do much because of the nature of the coolants and the fact that the fuel is a liquid and becomes inert as soon as it gets exposed to lower temperatures.

Let's be reasonable, don't you think the agencies and security forces don't factor in these possible risks and anticipate them? What about inherently safe facility design. What about the steep learning curve after each nuclear accident like Fukushima and Chernobyl?

Science a la carte

If you feel that terrorism is a threat, why not simulate an attack, try to get into one of those places unnoticed and tell me how it goes. It's really ridiculous and there's always police and army standby and on alert might some fool try to do something.

There's also the "potential for a catastrophic accident" argument. The same one I encounter in any of the anti nuke's long yet completely and perpetually regurgitated lists of concerns.

The funny thing is that even if you address all the questions they raise pertaining nuclear energy, and you propose new designs that help increase passive safety, decrease nuclear byproducts, and increase efficiencies; even then people like Helen Caldicott will find some reason to oppose it. Consider Caldicott to be a dogmatic person, a true demagogue, someone with immense sunk costs, no amount of falsification of their warped theories will cause them to change their mind, which isn't the purpose of this book but I suspect that you already know this. I want to open _your_ eyes and engage your critical faculties.

Suppose we have a reactor design that does the following, utilize spent fuel and bomb material to create electricity, and thus literally eats nuclear weapons and nuclear waste, and does it safely and reliably, would you oppose it? They would! If they were intellectually honest they would be pursuing the technologies that could solve their "thousands of years of problems and impending doom" arguments. Yet they present these perpetual catch22's and never ever acknowledge any progressive arguments.

The ultimate form of dogmatism is the pertinent denial of the fact that nuclear energy can innovate and improve significantly. Appeals to technological advancements are constantly made with regard to wind, solar and storage, but when we say that nuclear power can improve ten to a hundred fold, they simply shake their head and say no, they deny it outright. There are absolutely no redeeming qualities in Nuclear if these people are concerned, and this leads to the endless muttering of mantras and the show of extreme forms of demagoguery.

— Friends of the Earth

Simply consider their policy statements on nuclear energy.

No-nuclear justification:

"No-nuclear pathways are certainly technically possible," says *Professor Dave Mackay, former Government Chief Scientific Officer in the Department of Energy and Climate Change (DECC)."*—Friends of the Earth nuclear policy PDF.

Nuclear waste is dangerous:

"Nuclear power produces waste that is dangerous for thousands of years."—Friends of the Earth nuclear policy PDF.

No-nuclear justification (note the challenging part):

"The world-leading climate energy and research institute, the Tyndall Centre at Manchester University, independently reviewed evidence for and against nuclear power.—It concluded that Friends of the Earth's non-nuclear approach is credible, although challenging "—Friends of the Earth nuclear policy PDF.

Renewables are even healthier than nuclear (citation and clarification please...):

"Although producing electricity from nuclear power is healthier than coal or gas, renewable energy combined with energy efficiency is a much healthier solution."—Friends of the Earth nuclear policy PDF.

Science a la carte

— Rabid anti-nukes

As if all of this isn't depressing enough, there are always extremes, even more extreme than the likes of Greenpeace for instance, and if you thought that impossible, guess again. There are individuals that have issued death threats to scientists and engineers and communicators that report positively about nuclear energy. "*Hang the nuke gang*" was one of their *slogans*.

One of the most prominent anti-nuclear and ideologically driven groups out there is Greenpeace. The Nazca Lines... Their anti-nuclear activism is a boon for fossil fuel companies because they know that stagnation of nuclear power perpetuates the use of coal, oil and gas. Especially gas is becoming the *get-out-of-jail-free-card* for these companies because it is accepted as a perfect bridging technology, even by some environmentalists. The closure of Vermont Yankee, which was championed by Bernie Sanders, has resulted in the construction of multiple gas-fired power plants, which basically means that we've now substituted zero carbon technology for carbon emitting technology. Quite a leap forward! —end of cynicism. The loss of nuclear capacity will always be offset by building either gas-fired or coal-fired or biomass power plants, even if a couple of hundred wind turbines would be build to recover the loss of nuclear generation capacity, alongside an equal amount of fossil-fired capacity would be build, simply to offset intermittency.

And that's what anti-nukes are achieving with their show of ideological numbskullery. Fortunately, these people are very rare albeit very vocal, and they still have some traction within certain communities. Remember cognitive dissonance?

The worth-nothing of ideology

— Bernie Sanders: Senator and almost ideal candidate for POTUS

I want to end this grueling, and way too long and tiring and gloomy litany with a *catch22*, something that keeps me at odds and this pertains the American Presidential Election of 2016. This particular subject would perhaps be more at ease in the "Numbskull Politics" chapter, but I've chosen to bring it over here, and in the end you will get why.

While I am writing this in March 2016 I am being torn between wanting to endorse Bernie Sanders and advising against it. As you all have come to know, I place high value in social enterprises, taking care of one and other and making sure that all human potential gets a chance to flourish. Bernie Sanders seems to be the ideal candidate to help the US progress into a state in which this becomes slightly more real. However, Bernie Sanders has been quite vocal regarding renewable energy as a means to mitigate the adverse influence the US has on the climate.

Is this campaign geared towards renewables exclusively or does he also acknowledge the worth of nuclear energy? Let's see what his personal campaign website tells us about nuclear energy.

*"Create a Clean-Energy Workforce of 10 million good-paying jobs by creating a 100% clean energy system. Transitioning toward a completely **nuclear-free** clean energy system for electricity, heating, and transportation is not only possible and affordable it will create millions of good jobs, clean up our air and water, and decrease our dependence on foreign oil."*

Here we see a small excerpt from berniesanders.com, he wants to transition to a completely nuclear-free clean energy system. This is strike one.

*"Begin a **moratorium on nuclear power plant license renewals** in the United States. Bernie **believes** that solar, wind, geothermal*

power and energy efficiency are proven and more cost-effective than nuclear – even without tax incentives – and that the toxic waste byproducts of nuclear plants are not worth the risks of the technology's benefit. Especially in light of lessons learned from Japan's Fukushima meltdown, Bernie has also raised questions about why the federal government invests billions into federal subsidies for the nuclear industry. We can have an affordable carbon-free, nuclear-free energy system and we must work for a safe, healthy future for all Americans."

Here we have a collection of buzzwords clearly geared towards creating [mis]trust. Each point individually has been rebutted earlier in this book so I won't bore you with these factoids again but do acknowledge that this severely limits my position to endorse him any further. Rather than being wise and thoughtful like he is on a lot of different subjects, he throws his critical faculties out of the window and starts sounding like a dogmatic anti-nuke. Has he tried to see the opposite side of the coin? It doesn't seem so. Once he believes—taking things on by faith, rather than through empirical principles—he shuts all his critical faculties down and starts driving home this populist message. His opposition to the Vermont Yankee nuclear power plant, for instance, has caused carbon emissions to go up, rather than down, which ironically is counterproductive to his own goals.

There has been talk about shutting down Diablo Canyon—a nuclear facility in California. Consider the sheer volume of carbon-emitting energy it displaces simply by being put to work! It's there, it's staffed, it's being maintained, and it operates without fault. If this moratorium on license renewals proposed by Bernie Sanders becomes a reality, and the license for Diablo Canyon wouldn't be renewed or revoked, this nuclear power plant will probably never operate ever again, an incredible waste. Once again we are forced to conclude that this very dogmatic course Sanders has taken actually is counterproductive and doesn't lead to the goal he has set for himself: alleviate our negative effects on Climate Change.

Sanders wants to get elected, and he promises to invest time and effort into making these anti-nuclear statements a reality, whether he will be able to or not is irrelevant at this point. You always have to presume that a candidate will keep his word otherwise the candidate will be cheating his mandate.

It pains me that I have to state these issues but truth will out, how painful it may be. Rather than imposing a moratorium on re-issuing licenses, Sanders should acknowledge that the well-maintained reactors operating in the US are a big force for good in the world, especially when carbon-free / clean-air energy is concerned. Sanders should also take note of the intellectual and technological fortitude that exists within the US, it harbors the power required to make nuclear technologies leap forwards. I love to watch Isaac Asimov Memorial Debates and several other scientifically oriented videos and people like Neil DeGrasse Tyson keep driving home the message that we need to be doing more science and opening up paths to more science. I would love for the US to become a frontrunner in development and innovation again, and nuclear energy is one of the most potent areas in which the US may progress. But it can't and it won't if Sanders would become president and as long as his term would last.

The final nail in the coffin: "*When it comes to taxpayer support for nuclear power there is no end in sight.*" Sanders said Thursday, March 15, 2012. This is clear and leaves no room for interpretation. But I will rephrase his sentence to show you how flimsy it is:

"*When it comes to taxpayer support for renewable power there is no end in sight.*"

"*When it comes to taxpayer support for fossil-fueled power there is no end in sight.*"

Science a la carte

It's quite simple, this sentence is fit to use for any source of energy, and it even gets worse! If you look at the subsidies—worldwide—on a per unit of generation basis then nuclear is by far the best performer. Granted, it may have received a lot of subsidies, but it has performed exemplary in terms of cost per unit of generation delivered. There are no alternatives that even reach the TWh/$subsidy ratio. here Bernie Sanders is making himself unelectable to me. There will be a large contingent of smart, liberal and young voters that could help Bernie rise to power and I would find it a great move in social terms, but looking at energy matters, it could be disastrous. In fact, I would find it highly immoral to keep the US nuclear industry stagnant for the duration of at least one term.

Instead, Bernie Sanders should acknowledge the carbon-free nature of nuclear energy, the superiority of nuclear over any other kind of energy generation, including renewable (which is farming energy). With the advent of generation IV MSRs like the Thorcon reactor, the LFTR, and the TAP-reactor the US has an excellent chance to become a frontrunner in cutting-edge innovation.

Intellectual honesty and integrity demand of me to address this issue and it really stirs my mind. This is a true catch22; I simply cannot endorse my favorite candidate, because he opposes one of the most credible solutions to climate change, the thing he says he wants to fight. To me, it seems paradoxical. And if I could choose between abundant energy and a temporary president, I would choose the technology, because that will have bigger implications all across the world.

What are the alternatives? Hillary Clinton will probably mean a perpetuation of the status quo in terms of energy, no real social breakthroughs but at least it wouldn't set the US spiraling into societal collapse which could happen if one of the Republican candidates gets elected. She might be good in terms of reviving nuclear innovation, which is my hope. However, The ultimate opposition in the US at this moment is simply too wide and leaves little wriggle room for mistakes, electing the wrong candidate could prove disastrous.

Consider this metaphor, if you have the risk of a meltdown measured in (arbitrarily) 1 in 1000 years or the effects of climate change and you put them on a scale, there's no way of getting equilibrium, you can add one hundred thousand nuclear reactors and the scale would remain down on the side of climate change, it simply is too big of a problem. There's no way to justify not mitigating the effects of climate change. And I also have to add that the risk of meltdown is decreasing rapidly and steadily and with the advent of newer and even *better* technologies will disappear completely. Note the use of the word "even", it doesn't mean that contemporary reactors are bad, it simply means that we can still progress—a lot, and should.

I hope this chapter has made clear that those ideologies that include fallacious reasoning, and transform these into untouchable doctrine, are inherently dangerous and keep us from making the correct and most well-informed choices. Cognitive dissonance often smoothes critical faculties, it removes doubt and adds fervor. Driven by these corrupted ideologies people will make crucial mistakes, and some have the potentiality of being disastrous. Consider this possibility, regardless of being a left-leaning liberal myself, I suspect that people are probably going to question my motives of addressing Sander's anti-nuclear stance, up to the point of Ad Hominem. I work to free myself from group behavior continuously, and being intellectually honest to myself is paramount, so should it be for you.

I hope that Sanders will change his mind when confronted with the facts and that ideology and dogma will not push policy into the wrong direction if he gets elected. But he does get the wrong impression from those who believe what Mark Z. Jacobson and the suchlike believe, And this—misguided or ignorant or disingenuous — movement includes notable and well-known individuals like Joseph Romm, Jim Green, Naomi Oreskes, Bill Nye, Bill Mckibben, Naomi Klein, and last but certainly not least celebrities like Leonardo Di Caprio.

I submit to you that these so-called climate-change-experts and 100%-renewable-experts and [anti-] nuclear experts are being counter-productive to their own goals. In fact, I will assert that some of these people are willing to exacerbate the effects of climate change, simply by dogmatic opposition towards nuclear energy. Some of these people are raving lunatics, while others might be misguided or simply disingenuous. Whichever it is: Do not follow blindly and question and challenge everything! Even when your most popular star presents a view of the world that looks too good to be true.

It's the faith and the a-la-carte-argument... You can't choose credible climate change solutions by faith; these are forced upon us by reality. I wonder if these people really are as credulous as they seem, can they actually substantiate their 100% Renewable claims, or do they really take this stance on by faith?

We have to accept the truth about some cognitive processes that make it easier for some people to accept certain notions about reality albeit erroneous.

Up until now I have talked with many friends, and people I meet daily and I have yet to encounter someone that I cannot convince of the merit of nuclear energy. In fact, consider for instance my run-ins with anti-nuclear people on Twitter, even they become silent after a while. Not because they have run out of arguments but because I present mine in alignment with theirs. If they are honest and really care about mitigating the effects of the combustion economy, they will eventually have to concede that nuclear energy is one, if not the solution to anthropogenic climate change. Most of these people simply fold and become silent, probably because of sunk costs, and probably because they have surrounded themselves by people that are heavily invested in the erroneous idea that wind and solar are going to solve our energy problems. They might have vested interests and have companies or other contractual obligations that keep their households afloat financially. In this case, I ruefully admit that I can sympathize.

The other option is that they simply ignore me further and keep going on, the true demagogues. Even though I might sympathize with the first category, I do think they should be won over eventually, despite the possible ramifications. There is no reason to keep the ruse up if you want to save the planet, and you know that you've chosen a path that doesn't facilitate this goal then you should reconsider regardless of the possibility of sunk costs. We need the energy equivalent of Daniel Dennett's "The Clergy Project", an organization that helps these people transition to different fields of work. I would hate to see their talents go to waste on the implementation of meaningless things.

Simon Sinek is one of the people who has the tools in his box to make perfect sense of this chapter. It's the "EDSO"-argument. The what? Biochemistry: Endorphins and Dopamine and Serotonin and Oxytocin. These chemicals play a large role in our day to day dealings, they also determine things like trust, if people trust you they are more likely to hear a dissenting view from you and perhaps accept or adopt it as their own. I suggest you look up any of Simon Sinek's books or talks, he has done some particularly interesting appearances at TED conventions.

— www.startwithwhy.com

"Why leaders eat last", "Start with why - How great leaders inspire action"

Science a la carte

Robert Stone, the creator "Pandora's Promise", and an ex-anti-nuclear activist also expands upon this notion by explaining that if you want to convey a message, it is about alignment. Suppose you are a chess player, and you are a member of a certain chess-club, then people that are also members have an alignment with you, they share the same club, and there you have a chance to share ideas that might otherwise be rejected out of hand. Why? Because they probably are friendly with you because of the shared interest for the game and the membership of the club.

Consider for instance the different activisms that help me create lots of different contacts. I am an Anti-Theist (if you do not know what that is, look up Christopher Hitchens) and as such I engage in discussions about theocracy, religion, secularism, and make a lot of friends that are also interested in those subjects. As such I have an alignment with these people, we're all aligned because we call ourselves anti-theists, and we acknowledge the worth of questioning religious practices and beliefs. At some point, I engage with these online friends in order to determine what their stances on climate change and energy are and more often than not I conclude that these people are very open to my arguments and accept the premise that more nuclear power is necessary and that we can improve upon our past nuclear practices. I was immensely surprised when the owner of a bookstore, that is dedicated to free thinking, reason and science, read my first book and let me know that he agreed with almost everything I've written while I anticipated that some of the subjects might be considered taboo.

There are some who have developed the same cross-subject alignments as me. For me, I always tend to cross over from science to technology and from politics to economics all within the confines of the Climate Change discussion. This becomes evident when you read my first book *Highway to Dystopia.*

I do not go about proselytizing nuclear energy, but I do want to raise awareness, and that's why I engage in these kinds of exercises with everyone, and that's why I write these books. You should try to *spread the message* once you've examined the evidence and have concluded what needs to be done. It is essential because there's too much cognitive dissonance about which needs to be taken away.

The 100% WWS fallacy

There have been numerous people that presented farcical 100% Renewable strategies. Consider for instance the "Global Apollo Programme" written by the highly estimable Sir David King, Lord John Browne, Lord Richard Layard, Lord Gus O'Donnell, Lord Martin Rees, Lord Nicholas Stern, and Lord Adair Turner.

Yes, that's right, Lord's and Sirs. And take a look at their collective work experience, quite impressive:

Former U.K. Government Chief Scientific Adviser, Executive Chairman at L1 Energy. Former President of the Royal Academy of Engineering and former CEO, BP, Director of Wellbeing Programme, LSE Centre for Economic Performance. Emeritus Professor of Economics, Chairman, Frontier Economics. Former UK Cabinet Secretary. Astronomer Royal and former President of the Royal Society and Master of Trinity College, Cambridge. Emeritus Professor of Cosmology and Astrophysics. IG Patel Professor of Economics and Government, LSE. Chair of the Grantham Research Institute on Climate Change and the Environment. Senior Research Fellow, Institute for New Economic Thinking. Former Chairman of the Financial Services Authority and of the Committee on Climate Change.

It is funny, though, I have written my rebuttal to their Global Apollo Programme quite a while ago and yet I don't find it interesting enough to address now. We've got an even better 100% Renewable—or Wind, Water, Sun / WWS—scenario to dissect.

The 100% WWS fallacy

— Mark Z. Jacobson: the go-to guy for renewable fantasies

I've already spent countless of hours writing rebuttals to Mark's ludicrous claims, and I will extract stuff from my blog because it has been written for publication in the first place. Why Mark Z. Jacobson? He is someone who is trying to provide academic credibility to 100% renewable theses. In fact, he is the "go-to" guy if you want to show that 100% renewable—or in his case 100% WWS—can be done. I think he is wrong; I have examined his work and the physics behind our energy-consumption economy. I will point out the mistakes in his line of thought, he, on the other hand, dismisses me as an advocate (yes we have had a personal exchange of words). If I'm an advocate, he's an advocate disguised as an academic... Can you spot the difference? At least I am honest enough about it and need not use the Argument from Authority fallacy.

I called him out on Twitter, he answered, and I pinned him to the wall and subsequently he copped out by calling me an advocate, such foolish nonsense from a person who professes to be an academic. It is time the *demasqué* happens, to end this preposterous pseudo-academic nonsense once and for all. Mark Z. Jacobson is a professor, no less, at Stanford University, no less...

At first I was enamored by the solutions project. I e-mailed them because I wanted to know more, no response... Then my critical faculties kicked in: "*Is this the only man providing scientific evidence, is he going to determine what future my children are going to live in? Is he right? My gut tells me no! Let's find out what's going on here.*"

I have always been fascinated by skepticism and science; it gives us the opportunity to shed some light on the questions nature imposes on us. By nature, I mean the whole spectrum of Physics, Chemistry, Mathematics, Geology, Biology, Cosmology, and so forth.

Science a la carte

One of the videos anyone interested in science and skepticism needs to see is the one about the "Baloney Detection Kit" by Michael Shermer. One may note that the Baloney Detection Kit actually was a construct by Carl Sagan "*The fine art of Baloney Detection.*" Essentially it is the acknowledgment that if any one of these tens points gets assessed to be highly doubtful, we can call it baloney. It doesn't matter if the claimant is a regular bloke off the street or a Nobel Laureate. Academic pedigree or acclaim does not matter. If there is a truth claim, you can use the baloney detection kit.

What about the 100% WWS schemes such as the "solutions project" by Mark Z. Jacobson and friends. Note that the 100%WWS scheme not only tries to prove the feasibility of the 100%WWS future, it also tries to prove that there is a necessity to stop all activities connected to nuclear energy. This is the prime reason why I have chosen to *single* Mark out, he's not alone, but he doesn't mask his anti-nuclear activism—which is a word he uses to disqualify his opponents. Also, his 100%WWS scheme is being taken far too serious.

1. How reliable is the source of the claim?

Strike one: All the evidence presented concerning nuclear energy, slant to one direction. Jacobson **only** provides negative evidence, he doesn't acknowledge the positive evidence that has been provided by decades of operation and therefore negates the argument that nuclear has no role to play in a non-carbon-emitting energy landscape. Each time Jacobson is pressed on this issue, he falls back on the strategy of rhetoric and solely relies on portraying negative issues and resorting to logical fallacies such as ad hominem attacks or arguments from authority.

He does this in order to reinforce the necessity of a rampant growth model for WWS.

Strike two: he doesn't look at resource limitations well enough and doesn't acknowledge what needs to be done in order to ramp up production figures.

Suppose we would make a simple matrix showing positive versus negative arguments, it would look something like this:

My arguments in favor of nuclear energy

- High energy density
- Low material requirements
- Low [high value] waste production
- provides base load
- Ensures isotopic production for space exploration
- Ensures isotopic production for medicine
- Boon for chemists
- Good safety record
- Well regulated
- Long Lifespan
- Great steps forward in terms of cumulative energy addition per new unit

My arguments against Nuclear Energy

- Proliferation risk (which will decline with new tech.)
- Non-transparent costs
- Delays in construction due to regulations
- What if - angst for catastrophic events
- Requires Uranium and/or Thorium mining
- High-security requirements

Do you see what I did here? I don't cherry pick arguments; I present both positive and negative arguments regarding nuclear energy. Also, acknowledge that the arguments against nuclear energy are all fixable. However, Jacobson's argumentation looks like this:

- Proliferation risk
- Non-transparent costs
- Delays in construction due to regulations
- Requires a grid
- What if - angst for catastrophic events
- Fear of radiation
- Requires Uranium and/or Thorium mining
- High-security requirements
- Loss of land due to waste storage
- Long-lived waste products

Note that these are the ones he states, there are many more arguments to make regarding these technologies, I have never seen anything said by Jacobson that was positive with regard to nuclear energy, and I am not convinced that he will, unless he recognizes and acknowledges the flaws in his own research but the sunk costs argument probably keeps this from happening.

Which one of these is more honest? The one that weighs positives and negatives, or the one that ignores that there are positive points to be weighed? What do we learn from this? Not only is Jacobson *not* trying to prove himself wrong, but he's also trying to discredit other technologies by providing negative arguments exclusively, and this supposedly strengthens his position. This is a fundamental treachery to Jacobson's academic credibility. As an academic he should also take note of things that are contradictory to his claims, but does he?

Is it reasonable to expect that nuclear innovation will amend those negative bullet points? I think it is, and later in this book, I will present to you the reasons why I think we can solve all issues connected to nuclear energy as it is known today.

Also note that evidence in this respect is rather subjective, especially in matters where the jury is still out, and no certainty can be claimed and furthermore as soon as fear is a factor an appeal to credulity and irrationality is made, which is practically the only logical outcome of the "terrorists, bombs, proliferation, and long-lived waste" arguments.

2. Does the source make similar claims?

I question whether Jacobson is susceptible to positive claims / evidence in favor of nuclear energy because it flies in the face of what he has been telling everyone for so long. In fact, he is one of those people who try to pit the influence of the public against nuclear technology. This is called the sunk-cost fallacy. And it shows because he tries to present his own hypothesis based on overly optimistic possibility biases while he does the reverse for nuclear energy, he makes them diametrically opposed to each other. While it is not necessarily so, we need both technologies on the peak of their ability, but at some point, WWS/RE needs to be turned down because when its cumulative requirements kick in, we're in for sunk-costs ourselves, diminishing returns follow quite quickly after you commit fully to wind and solar.

What if the MSR revival arrives into the practical testing stage and proves that nuclear energy can be done for the same or lesser price than renewables; is almost completely proliferation resistant; meltdown proof; walk-away safe; and works at atmospheric pressures rather than High-pressures, would he then change his mind? Would he then do what any intellectually honest person would do? Say that this is much better than he thought? Instead, he intends to shut down all possible pathways of innovation and progress, if he gets his way, we will never know.

3. Have the claims been verified by somebody else?

The feasibility study has been reviewed by peers, so the answer to this question is: Yes.

104

Science a la carte

The question is who has done the review? To what extent have they tried to disprove the 100% WWS Hypothesis *and* disprove the part of this hypothesis in which nuclear energy gets addressed. What if the claims made about nuclear energy are not substantiated well enough, what does this do to this feasibility study? Does it then take away grounds for this study in the first place? Wouldn't it be more prudent not to talk about nuclear at all? What about vital omissions such as cumulative upkeep and replacements?

I question the validity of this peer review because of the numerous facts that have been omitted and have not been addressed, yet are essential.

Secondly, note that the peer-review process isn't perfect as can be read in an article called "*Peer review: a flawed process at the heart of science and journals*" which has been published in the Journal of the Royal Society of Medicine. Take note that I know academia very well, I may not have attended university or college myself, but family members have, friends have, and it is not as good or as rational and as reasonable as it should be. In fact, there's a lot of bad science going on today. One may note that peer-review doesn't always filter out errors, and doesn't always provide alternative views, which probably has happened in this feasibility study. And I wish that it was different. That being said, peer-review is also the best process we have for validating research findings, so it is a double-edged knife. We always have to be prudent and be prepared to question the findings and the validation thereof—remember the "faster than light neutrino" controversy? Google it...

Since Jacobson grounds his articles in subjects that are heavily influenced by biases, I question the validity of his findings. Why for instance have we no full analysis on material restrictions, production restrictions. There are some, but they are incomplete. Also, consider that Jacobson bases his PV scenarios on one particular brand and type of PV technology, what are the material requirements for this particular technology per unit? What kind of chemistry is required? What kind of rare-earth materials? I don't know, neither does he, at least it doesn't show in his papers. These are vital omissions because they are fundamental to the feasibility of his thesis.

Also, take note of an article that is named: "*How Many Scientists Fabricate and Falsify Research? A Systematic Review and Meta-Analysis of Survey Data.*"

"*and up to one-third admitted a variety of other questionable research practices including "dropping data points based on a gut feeling", and "changing the design, methodology or results of a study in response to pressures from a **funding source**.*""

There are many academics that like to point out that Jacobson's studies are suspect because of questionable sources of funding and possible double agenda's. I'm not inclined to dismiss or to accept these off hand, but they give one pause. Especially when confronted with a very defensive and dismissive Jacobson when pressed.

4. Does this fit with the way the world works?

The answer is no, Jacobson blatantly disregards resource limitations, even though he has done a study on resources, it's fallacious, just as this one is. In fact, if you stack studies in this manner, you're bound to see all of your work collapse if one of the cards gets pulled out. In order to reach his goals current [world] production levels of wind and solar have to be increased by at least fifteen times. This requires equal increases in mining, transportation, and purification processes. Not to mention the amount of denudation required to get these resources, if we can get them, which is highly questionable.

Also, consider the fact that he wants to shut down the entire nuclear industry, this he suggests in his article "*100% Clean and Renewable Wind, Water, and Sunlight (WWS) All - Sector Energy Roadmaps for 139 Countries of the World*", which is the basis for this rebuttal. This means that we will shut down all flux-reactors required to create medical isotopes, research isotopes, isotopes required for space exploration, etc. Has Jacobson ever been to the radiology department in a hospital? Has he ever had a dental X-ray? No? It's time he gets educated on the ancillary benefits of the nuclear industry, which are certainly not negligible. One may also note that Cobalt-60, for instance, is being created in Canadian CANDU reactors that create both electricity and special isotopes.

Consider this statement by Jacobson: "*The roadmaps presented here assume the adoption of new energy - efficiency measures, but they exclude the use of nuclear power, coal with carbon capture, liquid or solid biofuels, or natural gas because all result in more air pollution and climate - relevant emissions than do WWS technologies*"

Which basically means "*There are more emissions tied to nuclear as there are to renewables*", which according to reports from the EIA and IPCC is incorrect. But then there's this "*in addition to other issues, as discussed in Jacobson and Delucchi (2011) and Jacobson et al. (2013).*" This adds to the ambiguity... Does this mean that nuclear indeed emits more than renewables according to Jacobson, or does it merely mean that nuclear is part of the *additional issues* yet doesn't emit as much? If the claim is indeed that nuclear does emit more, he has already been proven wrong by numerous academics and institutes and organizations from all over the world, reputable ones...

Take note that he really wants to shut down the entire nuclear industry as can be seen in "*Table 10, Estimated 139 country job losses due to eliminating energy generation and use from the fossil fuel and nuclear sectors.*"

This means shutting down everything from mining to energy production but also the production of medical isotopes and isotopes for a great host of other activities.

The effects of our activities on the planet have forced us to show our hand. How will we solve the direst threats of ocean acidification and the ever increasing demand for fresh water? The reality is that we cannot discount any non-carbon-emitting energy source. We will see an increase in energy demand because of a growing world population. We will need more energy than ever before in order to capture carbon and sequester it in basalt. We will also need substantially more energy to build sustainable water networks that bring water to drought-stricken areas. Recharging aquifers and defending areas against the effects of desertification is going to be paramount.

ERGO

The slow and gradual and eventually stagnating pace of the 100% WWS scheme proposed by Mark Z Jacobson will not satiate the growing demand for energy.

To make matters worse, Jacobson fully omits cumulative upkeep and replacement requirements for wind and solar, which range between ten and twenty-five years, which will require annual additions to grow exponentially, and eventually make sure that the growth curve bends downwards again. Also, consider that even though many renewable proponents claim that renewable energy is "growing fast" on the larger scheme of things, this is absolutely meaningless. Renewables (without almost maxed out hydro) is less than a percent of the total energy consumption in the world thus totally negating the proportional "large push" of renewable energy sources. 50% of 1% is still only 0.5%. Even if it would grow by 100% in the first few years, it wouldn't amount to much. In fact, a growth ratio of 1500% is required, which makes the task at hand a rather steep one—and, dare I say it, insurmountable.

Also, consider the fact that each 100% WWS / RE scenario depends on an increase in the size of grids, an increase of the complexity of grids, and a vast array of energy storage devices, which will compound the problems even more. It makes grids more complex, not less, it makes grids grow larger, not smaller.

And then there's this, has he accounted for increased water desalination and management requirements? Has he accounted for increased energy use in agriculture? Has he accounted for carbon capture and sequestration?

Why ask these questions?

To maintain human civilization and to maintain stability we need large volumes water and food. Both necessities are being stressed tremendously. Shortages of water and food lead to famine, instability, war and ultimately death. We need to increase water management and infrastructure to counter the ever-changing precipitation patterns; we need to increase water management to keep people healthy and fed. We are overdrawing fresh water sources, we're outpacing the replenishment raters, and to make matters worse our alternative is to produce brine in the process of creating potable water. We've got a significant problem on our hands, that is going to require a tremendous amount energy. Also, consider the fact that the Ocean is a storehouse of food for roughly one fifth of humanity; once this oceanic food pyramid collapses it will have severe repercussions. Ask yourself this question: "what will happen if Sardine populations diminish completely?" And see what scientists have to say about it and start by looking at Plankton and its vulnerability to acidifying oceans. Or consider the future of the bees, the alpha-pollinators, the most important insects that make sure that most of our crops and fruits can grow and keep us nourished. The food baskets of the world are at great risk from droughts or torrential rains, increased heat, unmanaged pests due to softer winters, and the possible demise of the quintessential pollinators.

To counter these issues we need energy, far more than Mark Z. Jacobson dares to admit, because once he acknowledges this his house of cards will collapse. I will not stand idly by while these people are peddling nonsense and keep feeding it to the gullible.

5. Has anyone tried to disprove the claim?

Countless people **have disproved** his claim. The fact that James Hansen and a plethora of other scientists claim that is not possible to save the world using only renewables/WWS is a testament to the fact that Jacobson's findings are probably invalid. Also, note that when confronted with claims such as *"Nuclear energy is also required to fight anthropogenic carbon emissions"*, Jacobson and his colleagues resort to name-calling and attempts at character assassination. Jacobson did it to my face, I still have the tweet as proof that he did it, he called James Hansen's views *extremist*, what he meant by it? I don't care... Add context if required.

6. Where does the preponderance of evidence point?

The first acknowledgment we need to make here is that he doesn't set out to disprove the 100%WWS. He is looking whether it can be done, that's why he has called it a feasibility study. As such he sets out to prove the 100% WWS hypothesis, without looking at true limitations. Past and current production rates of wind-turbines support the view that it is hard to increase production significantly, at least not by the rates required to be meaningful and confirmatory of Jacobson's claims. Again he presents an overly optimistic scenario that slants away from what is really happening and what is possible. First of all, we may note that the growth of wind additions is increasing, there is, however, no reason to suspect that the growth will remain exponential. The evidence provided by the EIA actually refutes the notion that wind will remain growing exponentially. In fact, we may expect the growth curve to be curbed at one point. If we take a look at Germany's faltering Energiewende we may note that it's 1.1 trillion dollar wind-excursion has failed and will be shut down in 2019 putting an end to the pervasive myth that Germany is the shining example of renewable growth, the substantiation of which can be read in the Berliner Zeitung: "*Die Bundesregierung legt bei Energiewende den Rückwärtsgang ein.*"

Also, consider the fact that the US in World-War II was able to produce about 32 000 heavy bombers airplanes in a 4,5 year time span. There are people who claim that if we could do this, we can build enough wind turbines to help us progress to a future of carbon-free energy generation, this is preposterous! Mind you they couldn't get enough of them; of a grand total 260 000 airplanes built 95 000 didn't make it... We currently produce about 20 000 2,5MW wind turbines, which already is a larger total volume of materials than the bombers that were produced by the US in WWII. By Jacobson's reckoning, we would 1.95 million 5 MW units or 3,91 million 2,5 MW units, which all should be built in a timeframe of 34 years. This means that we would need to build 115 000 2,5 MW units per year (omitting cumulative upkeep and replacements) which is almost six times more than we currently manufacture. If we keep going at the rate we are going, we won't get there until the year 2211. I have yet to see any plans of increasing manufacturing capabilities to facilitate such growth. Note that this is 32% of Jacobson's energy mix, the rest basically is Solar only, which is mired with even more production caps and resource troubles all across the board.

7. Is the claimant playing by the rules of science?

The funny thing is that I wanted the 100% WWS world to become a reality, which meant that I went out to figure out if this was possible. I wanted to do the same as Jacobson has done. Ask yourself this question: how is it possible that Jacobson thinks it is feasible, and this *uneducated* guy from the Pays-Bas doesn't. In fact, after I did the legwork in order to be able to derive some conclusions I changed my mind. Jacobson doesn't, and that's unscientific. In fact, he attacks nuclear energy almost dogmatically, totally disregarding *any* positive arguments in favor of the nuclear industry. He has all the hallmarks of a demagogue, a man that has rigidly held beliefs and is inflexible because of sunk costs and doesn't accept criticism and can only resort to meaningless mantras when pressed.

8. Is the claimant providing positive evidence?

On first glance, it looks like he is providing positive evidence for his claims, yet upon further investigation his claims are too optimistic, as shown in the aforementioned arguments. I suspect that Jacobson, therefore, needs to introduce a laden subject like Nuclear Energy to fix the bias of the person reading the article.

9. Does the new theory account for as many phenomena as the old theory?

There is no previous 100% WWS hypothesis; we're now talking about a 100%WWS / 0% nuclear dichotomy.

10. Are personal beliefs driving the claim?

First question: Why does Jacobson try to do a feasibility study on 100% WWS scenarios? Subsequently, why does he need to dismiss nuclear energy in order to push his 100% WWS scenario? What sticks behind it? If he is a true academic I doubt he would be as adamant to go into extremes: the 100%WWS/0% nuclear dichotomy.

Also, ask this question: Why is nuclear even mentioned in his feasibility study? Or why the entire nuclear industry (including the production of chemical isotopes) should be shut down?

If this is a feasibility study to prove the possibility of a 100% WSS future, why is it necessary to say that nuclear energy is not a credible option? Here's why: Jacobson has to wipe all other sources of energy from the table, even the well-regulated, safe and robust nuclear industry that provides a sturdy 3500 / 4000 TWh per year and has new reactors being built every year. I'll re-iterate this ad nauseam if need be, we need to address the issues of ocean acidification, a possible arctic methane release, and the ever growing water troubles on this planet. These three issues will drive energy demand upward, apart from the already exploding population. The 100%WWS scheme does not foresee to counter these three major impacts we have precipitated on the Earth. Without acknowledging these, Jacobson discredits his research even more. It is amazing to see a professor of "environmental engineering" disregard changing precipitation patterns and their repercussions.

Also, consider this fact: renewable energy looks good, it is the feel-good technology that kindles hope in the hearts of those who are afraid of the effects of anthropogenic climate change, but also those who glamor over independence and the end of the influence of *big companies*. And as such nuclear energy easily can be brushed aside. What will happen if we are going to produce the amount of PV panels proposed by Jacobson? Then the PV manufacturers will become the new big companies... We would be substituting one big industry for another. There are a plethora of 100% WWS / RE schemes out there since it caters to the hopes of people. Nuclear on the other hand only spells *doom and gloom* in the mind of many, but the question is this: Would we rather not build the most effective technologies to fight anthropogenic climate change, before it is too late? Or are we going to rely on what-if scenarios that depend on overly optimistic models, while discarding an entire non-carbon-emitting industry that can power a significant (note significant, not [yet] 100%) portion of our civilization?

This is the final nail in the coffin: James Hansen has always tried to explain what is happening to our planet and which processes are having adverse effects. He was one of the very first to do it. Now he steps up to annunciate that nuclear has to be part of the solution, people like Jacobson, Romm, Oreskes and many more jump onto the barricades to call foul. What drives this apparent nigh pavlovian knee-jerk response? Is Mark Z. Jacobson really interested in getting humanity out of trouble as fast as possible? I doubt it. To me, it seems as if he is presenting feel-good-scenarios based on faith-based claims that are only partially substantiated but mostly based on unreasonable assumptions. And it seems that he is not using the scientific method correctly in addressing these issues.

The man is an academic, he is a teacher, he has a profound influence on students and the way they present their ideas and the way they quantify their research and what they can and cannot do. As such he should be held accountable. Also, he is someone who testifies before congress, and as such he should be held accountable, for he is trying to convince representatives to sink their costs into a cul-de-sac exclusively lined with renewables. I don't see any compatibility between 100% WWS scenarios and the dire plight against anthropogenic climate change and a human civilization that progresses.

Let's not mention Precourt's involvement in all this and its stakes in fossil fuels or the other way around, whichever way it is, this taints

Jacobson's work, whether he likes it, or not. After all, he is a senior fellow at the Precourt Institute.

Since I find fault everywhere, I will call Jacobson's 100% WSS scheme: **<u>BALONEY</u>**

Anticipating the apologia argument

"We choose to go to the moon. We choose to go to the moon in this decade and do the other things, not because they are easy, but because they are hard, because that goal will serve to organize and measure the best of our energies and skills, because that challenge is one that we are willing to accept, one we are unwilling to postpone, and one which we intend to win, and the others, too."

- JFK at Rice Stadium, September 12th, 1962

What is the purpose of this chapter? I wanted to show you how the debates between pro-nukes and anti-nukes go. Sometimes I feel as if I am engaged in nuclear-apologetics. The term apologetics crosses over from religious debates in which hardline believers are called apologists for the horrendous stuff that happens within the confines of religion. I want to show you however that nuclear energy has nothing to apologize for; in fact, it is a great force for good in the world. We will find out what arguments are commonly used against nuclear energy and how I make it a sport to counter them in such a way that the *spectator* becomes interested enough to go and find out what is true and what is not. I won't be trying to change the minds of those whom I am debating.

I won't have it said of me that I am immoral for being an advocate for the advancement of nuclear energy. I will be quite clear and if need be, even rude on this subject because arriving at this point has cost me a lot of my precious time, mental gymnastics and tireless hours of intellectual mentation. Consider this bold statement: The people in the 100% wind and water and solar club of clowns are not interested in trying to solve the issues at hand, they are knowingly pushing credible solutions out of the way and substituting them with their own foolish fantasies and open-ended experiments. They either do it knowing that it isn't possible to realize these ideas in which case these sick sociopaths deserve to be put in a sanitarium because they are willing to gamble with the survival of our collective civilization and the countless of beings that share this earth with us, *or* they are really as inept and incompetent and as dumb as they seem and do not deserve to hold any position of influence or power over others, especially those working in universities or newspapers or the government... (The accountability argument)

Each time I engage with people that are convinced of the 100% Renewable / WWS future, I present facts and figures, oftentimes I am being met with the same rhetoric which always goes around like this in random order: waste, cost, Chernoshima, radiation, terrorists, bombs. That's it; do these people honestly believe that these issues haven't been thought about by protagonists for nuclear energy? Debunking these arguments time and time again is tiring, but I will do it nonetheless. I am not putting in all this effort for the odd chance that I might turn my antagonist around, no... I keep doing this because there are onlookers to convince. Most of these onlookers—not all—are fair enough to let the evidence play a big role in forming their opinions, and this makes it worth the while. The sheer volume of skeptical people I've met and had great conversations with makes it worth the while.

It is a dumbfounding fact that the antagonists show this rigid frame of mind towards nuclear energy. These are people who try to envision a future in which our civilization is fueled by the power of the sun and the wind. This is an immensely difficult technological problem that requires far more innovation and progress than they care to admit and yet they think that the nuclear industry is a pool of stagnant archaic reactors with poor oversight and a lack of self-criticism. As such they propagate the notion that there are some unfixable issues with nuclear energy, which by itself is a very narrow-minded and dogmatic view to hold, especially when they fail to acknowledge the shortcomings of their own convictions.

Consider this strange juxtaposition of thoughts that roam around in the dark clouds that are suspended in their craniums: On one side they have no difficulty in envisioning a 100% WSS future while on the other hand they fail to see the possibility for innovations in the nuclear industry. While we see small increments that make renewable energies edge slowly in a forward direction, great steps can be made in the world of nuclear energy generation if we open up avenues of free inquiry and research and innovation. In fact, we have already seen technologies in the sixties that were orders of magnitude better than the LWR's and PWR's we see dominating the nuclear energy landscape today.

The more sensible thing to do would be to incorporate any and every non-carbon-emitting energy source to power civilization into the future. Consider these sources to be Fusion, Generation IV&III Fission, Geothermal, Hydro, Solar (particularly heat) and very limited wind (saving birdlife). My terseness mainly stems from the idea that these people want to exclude nuclear energy and think that they can run the world on solar, wind and very limited expansions on energy capture encapsulated in the hydrological cycle i.e. dams, wave devices, and tidal devices.

Let's go over these rhetorical issues, those that antagonists like Jacobson, Romm, Klein, Green and others deem unfixable or even insurmountable. I know that some of them become incredibly annoyed with me when I say it, but some of these arguments are unsubstantiated and narrow-minded and shortsighted. It's the double standard, the willingness to see unicorns running around in WWS land, but only trolls in nuclear land. It's the dogmatic and nigh compulsive juxtaposition that needs to be smashed into pieces. It is, in fact, irrational behavior. Amongst other things, the dread for nuclear energy stems from the apparent ominousness and misunderstanding of radiation. There are a plethora of total non-sequiturs and non-arguments and slippery slopes that I can spot coming from a mile away. Let's go over them:

"Nuclear energy has been stagnant, and old designs are still being used."

If anything has been learned from France's example it is this, any modern and industrialized country can and will be able to do a complete overhaul of its energy generation network. France started building their current fleet of nuclear reactors in the 1970's and two decades later managed to generate 75% of their electricity by using non-carbon emitting nuclear energy. France is also one of the leading nations in nuclear research, with CERN they have an incredible complex that is leading the world into unknown territories of particle-knowledge, and France also has 14 other research reactors which are used in materials and fuel and waste-processing research.

The reactors France uses for its energy generation have been standardized, they use three different pressurized water designs: 34 three-loop 900 MW reactors, 20 four-loop 1300 MW P4 type reactors, and finally 4 four-loop 1450 MW N4 type reactors.

France has also heralded its own Generation III+ reactor, AREVA's European Pressurized Water Reactor or EPR. One of which is being built near the Belgian Border in Flamanville and is expected to come online in 2018, another is being built in Olkiluoto —Finland, two more are being built in Taishan—China, and two have been proposed at Hinkley Point on the southern shores of England but has been mired in bureaucratic issues.

Anticipating the apologia argument

And then there's ITER (International Thermonuclear Experimental Reactor), one of the attempts to prove the feasibility of nuclear fusion, The world's biggest Tokamak is currently being assembled in southern France, the project is incredibly complex, and it is expected that in June 2016 a tentative schedule will be presented—after all, this is an experiment. Germany has its own fusion project called Wendelstein 7-X, it's a stellarator, Google it! It's an amazing piece of machinery which seems to be wrought from the mind of a genius, and probably is.

France is an excellent showcase of what can be done if one commits to a plan, and they have done so excellently, not only by committing to standardized designs but also by creating a sizeable research environment and a sense of optimism. But will this optimism last?

Let's not forget that France has the least carbon emissions per person (5.7 metric tons) of all of the strongly industrialized countries Europe trumping Germany (10), Denmark (7.1), Norway (8.9)—the so-called bastions of *green energy*. Far better is it to follow the examples of France, Sweden, and Switzerland who have embraced nuclear energy and unsurprisingly end up on the low-carbon per head end of the spectrum.

China is one of the countries that are investing in nuclear energy like gangbusters, smog is killing their inhabitants, and they know they will run out of coal, they know that hydro energy has been maxed out, and they also acknowledge that to keep growing and to keep progressing they will need more energy than ever before. What are they going to do?

They are building EPR's from France, AP1000's from the US and a myriad of different designs. They have developed a working pebble-bed reactor and are also working on breeder reactors and molten salt reactors. They are pushing as hard as they can. Anyone who is enamored with the idea that nuclear energy is a stagnant technology simply has to look around in the east. Azerbaijan, Georgia, Kazakhstan, Mongolia, Bangladesh, Sri Lanka, Indonesia, the Philippines, Vietnam, Thailand, Laos, Cambodia, Malaysia, Singapore, and Myanmar are all considering nuclear energy as a viable option to help their people progress into a new age of prosperity.

The list is actually longer, there are forty-five countries in total that have set out on a quest to become involved in the nuclear age. And even though some of these countries might be interested in generation II or generation III designs, it is hard to maintain that nothing is happening and that there's no interest and that there's no innovation going on.

One of the reasons that people might think this is because of the US, which, despite its technological fortitude, has mired nuclear energy in a cobweb of politics and bureaucracy. One may also note that there are experimental or pioneering projects that progress slowly and are beset with cost overruns and redesigns of individual components, and these are the issues that are magnified by the naysayers in order to keep the public unbalanced.

If you look past all these issues and take a glimpse through the slightly opened doors, you can see a great host of communicators, entrepreneurs, scientists, engineers—all academics—working hard to reach the next level in the nuclear age a possibility. Consider the work of the UC Berkeley nuclear engineering department or consider the research at other world-renowned hot-spots of research and knowledge such as the Massachusetts Institute of Technology (MIT) which has successfully trained two young and bright physicists who have started developing and commercializing a credible solution to the *nuclear waste problem*.

Anticipating the apologia argument

— Nuclear is unsustainable

By what measure? On fissile and fertile material alone we can reach thousands of years from resources on Earth alone. Not to mention reserves elsewhere in this solar system. If you consider Transatomic's design and acknowledge that there's enough energy in spent fuel and nuclear bombs to run civilization (the entire world) for at least seven decades and add to this the already unearthed thorium reserves (which are considered a byproduct of rare-earth mining i.e. considered unusable waste...), and we're already set for several hundred years. How is this not sustainable?

Not only do current reactors achieve fuel efficiencies lower than 5%, we're also burning the rare stuff, the Uranium 235 instead of the ubiquitous Uranium 238 (which constitutes 99% of the natural occurring Uranium).

It's a three step process forwards: 1. Start building and testing nuclear reactors that can run on Uranium 238, nuclear waste, spent fuel, and nuclear bomb material; 2. Start building and testing nuclear reactors that can breed Thorium; 3. Apply assembly line like construction processes to these reactor designs.

Also consider these basic figures that I got from the *September 2015 Quadrennial Energy report* by the Department of Energy of the United States: given its operational life cycle wind requires 22 times more materials to get build than nuclear over a 50 year time span, solar 36 times as much as nuclear per equivalent generation capacity, including the accounting of capacity factors.

Given the fact that the feedstock for new nuclear energy is already widely and readily available and the materials required to build the facilities are ubiquitous we may conclude that nuclear is even more sustainable than any other form of power generation / farming. And as such you may call this "nuclear is unsustainable" claim baloney.

— Nuclear energy is expensive

Almost always heard is the argument that nuclear energy is costly but simply because something is expensive upfront, doesn't mean it isn't a superior alternative. In fact, I submit to you that this is a complete non-sequitur. Why focus on cost, while you know that this is a nebulous issue, something that can be fixed as easily as flicking a switch? It shows the dishonesty because it is being presented as an insurmountable problem.

In this small section, I will try to make some sense of the economics of nuclear power plants and try to convince you that we've yet to reach the cheapest point at which we can build and run nuclear power plants.

Consider the IMSR design proposed by Terrestrial Energy, which estimates an LCOE (Levelized Cost of Energy—including everything from finances, feedstock costs, taxes, running costs, capital costs, decommissioning, etc.) of 40 to 50 Dollars per one MWh of electricity produced. It is also a matter of where the technology is being built, France for instance already has nuclear energy at 50 $/MWh, in France nuclear energy is partially government owned and the reactor facilities actually don't run during the weekends. In Great Britain nuclear costs somewhere between 80 and 105 GBP/MWh and only natural gas (55 / 130) and biomass (60 / 120) are cheaper. The same figures can be seen in the US where nuclear is only being trumped by natural gas, onshore wind, biomass, hydro and coal, but not by a very great margin I have to add. If you look at which technology is consistently more expensive we see that it is [any] solar to electricity system—and this is not taking into account the necessary *smart grids* and *storage* that are required in a 100% WWS scenario.

If we look at the figures provided by the National Renewable Energy Laboratory (NREL), we can see nuclear again playing a midfield position. The only possible conclusion we may reach is this, the "*nuclear is expensive*" argument is—again—baloney. Most of these people tend to focus on cost overruns during construction but completely forget that these reactors operate 50 to 60 years at capacity factors higher than 75%. And that's why the LCOE metric is so relevant. Once again the general public is being played by presenting only half the picture—large upfront costs—while reliability, life cycle, lifespan, and total running costs are completely ignored.

Also, take note of the predicted capacity factors—again by NREL—for 2025 where nuclear still trumps everything else at 90%, and PV and wind don't exceed 42% and 26%. Remember that capacity factor coupled with the nameplate capacity determines the actual output. When calculating things like LCOE we look at the cost per unit of energy generated over the entire life cycle of the technology. Given the fact that a nuclear facility runs for at least 50 years contrasted to about 20 years for wind and solar should make you think.

Furthermore, NREL expects the LCOE's for PV and wind and nuclear to be competitive with each other. How is that "nuclear is expensive" argument looking now? Especially when based on NREL figures. NREL is the US National Renewable Energy Laboratory. Also, take into consideration that NREL has not taken into account any other technology than the pervasive PWR and LWR designs and probably based the LCOE's on designs like the AP1000. Once the Thorcon MSR, the Terrestrial IMSR, and slightly later the FliBe LFTR and the Transatomic WAMSR hit the market the LCOE for nuclear is going to drop further, I almost dare to assert this with absolute certainty. We may also expect AP1000, APR1400, EPR and other designs to keep driving down the costs. The nuclear energy industry is set to revolutionize itself by creating more standardized designs that are easy to produce and quick to deploy because they have significantly decreased complexity, incorporate passive safety and are far more efficient than contemporary reactors.

Consider the Energy Information Agency's (EIA) view on nuclear energy: "*A nuclear power plant is a well-known technology that already performs close to its optimum performance. As such, EIA expects that capital expenditures will incrementally improve over time, slightly more quickly than inflation.*" Note that this citation is provided in an NREL report called "The annual technology baseline" from July 8th, 2015.

This shows that both EIA and NREL are assuming that nuclear power generation will remain as is. Perhaps based on AP1000 technology since they claim this to be "*advanced nuclear*" but we may well expect that fuel and generation efficiencies will climb with the implementation of different fuel types and mediums and the implementation of smaller yet more effective generators. As said earlier, despite the fact that PWR and LWR designs are quite mature, there are a thousand different and probably far more efficient ways of doing nuclear and therefore, it is entirely justified to assert that the learning curve in the nuclear industry will become much steeper as soon as we start building and testing nuclear reactors that operate on completely different principles. As such we may also conclude that nuclear energy is immature since we've only explored a dozen designs that all rely on solid fuel cycles. As far as I know, we have one proof of concept for a liquid fuel-cycle technology, and this was the successful yet, in the end, stifled MSRE (Molten Salt Reactor Experiment) at Oak Ridge National Laboratory.

Once we switch from Uranium 235 (0,7%) to Uranium 238 (99,3%) and Thorium and spent fuel (and bombs), we will see the costs for nuclear feedstock drop dramatically. We will seriously mitigate the cost for sequestration due to a severe reduction in existing waste stockpiles and new waste produced. The innovations that will make these great cost reductions possible are just around the corner, and the proof of concept has existed for decades—the MSRe at Oak Ridge National Laboratory. And it is still there...

"So what's the bottom line? That nuclear power is expensive only if a country chooses to make it so."—Can't remember who, but certainly not mine.

There you have it, I submit to you that the *"nuclear is expensive and ripe with cost-overruns"* argument is, in fact, a solvable non-argument that shouldn't be used at all. In fact, by raising this argument, you are willingly discarding a credible technological solution on nebulous terms. A total non-sequitur: Just because something looks or feels expensive, doesn't mean it is impossible to realize. Also, take into consideration that if you're faced with knowledgeable people and you try this argument it might explode in your face. After all, we can spout a lot of unsubstantiated nonsense without being held accountable. It is quite easy to browse the internet and find some document that claims that nuclear is the most expensive form of energy of all, but the truth of this claim can only be determined by the evidence. Always demand evidence and seek other reports that might contradict the claim. I am confident that I have seen enough evidence to the contrary.

—Generation III and Generation IV are a decade away

Generation III (or III+) reactors are the final iteration of the water-reactors we use today. These reactors feature passive safety systems. Passive safety means that no human interaction is required to keep the reactor cool for a limited amount of time. AP1000's and APR1400's and EPR's are considered Generation III+ reactors. Passive safety is possible because of simpler yet more logical designs. This simplicity also helps to bring the costs down.

Consider the APR1400—a pressurized water reactor—of which one went online in South Korea in March 2016, it has a capacity of

1400 MW and as such one may expect it to generate about 11 TWh of electricity each year. There are seven more being built and another four planned, which brings the total generation capacity per year up to 132TWh. This capacity should be added by 2022 with a unit coming online each year from now until then. The expected lifespan of an APR1400 unit is sixty years so until 2077 we may expect these reactors to produce a total 132TWh per year. Suppose you want to achieve the same feat with 2,5MW wind turbines. A total of 7920 TWh would need to be generated over sixty years. The lifespan of a wind turbine at current is barely 20 years (consider for instance the fact that the oldest wind turbine park in Canada is 23 years old and will not be refurbished, hence considered to have reached the end of its life-cycle) so you need at least three replacements to reach this 7920 TWh figure. A 2,5 MW, wind turbine at a capacity factor of 30%, will produce 0,1314 TWh in its operational lifespan, by this measure we would need to build 60 000 wind turbines over a 40-year time span to achieve the same effect as those 12 APR1400 reactors have during their operational life. Just to make things clear, we currently build about 20 000 of these wind turbines each year accounting for roughly 50 GW of total capacity.

There we have it, a new nuclear technology, not quite mature, being built for the first time, in a timeframe of six years adds 16.8 GW of capacity to the mix or 132 TWh per year equivalent whereas the 50 GW of wind capacity adds an equivalent 132 TWh per year. And take note that we're now talking about APR1400 versus wind turbines, I omit the 16 GW's of non-APR1400 capacity that will start being built in 2016 alone and there are a total of 165 nuclear reactors planned worldwide, and consider this to be a conservative number because even easier and cheaper designs already loom on the horizon.

Once standardized and simpler designs have been built, and the growing pains have been found and amended, and the learning curve has progressed, the deployment speed will rise, and the cost will drop. And this is not taking into consideration Terrestrial's and Thorcon's standardized MSR designs.

Then there's another Generation III(+) design of which there are currently eight under construction, the AP1000. The AP1000 is a generation III+ pressurized water reactor with a nameplate capacity of 1117 MW. Although these designs are delayed due to some necessary modifications, these are being implemented. This is part of the learning curve and will ensure higher deployment speeds in the future.

Note during the finalization process of this book China announced that they will be the first to have an operational AP1000: "*Chief Engineer of State Power Investment Corporation Wang Jun confirmed Wednesday that the world's first Westinghouse Electric Co. AP1000 reactor would be operational by the end of the year (2016)*."

— Chernobyl and Fukushima

Let us start with this acknowledgment: What blithering idiot builds a reactor based on a design known to be faulty, and even considered dangerous, and lets it being operated by unqualified operators and doesn't accept criticism or considers improvements and necessary modifications (Chernobyl).

Second acknowledgment: What blithering idiot builds a reactor in a region known for Tsunamis, builds a low sea-wall and then sticks all the important parts in the basement and forgets to secure the fuel required for backup cooling?

We should visit those responsible and hold them accountable. First with a firm smack to the head for the utter stupidity of their actions and the infliction of nigh irreparable damage to the credibility of the nuclear industry and the loss of three gigawatts of perfectly functioning / non-carbon emitting capacity in Japan. Secondly, we should offer them a "*thank-you*"—grudgingly—for precipitating the chance to experience these things and get a firm scientific grasp of its repercussions and most importantly, learn from it. Thirdly we should take a long and stern look at the existing nuclear industry and point the finger, "*don't you ever fuck up like that again*", not because of apocalyptic visions of doom and gloom (because they do not exist), but simply because it will—and has—cripple[d] our efforts to mitigate the effects of the combustion economy and we must not accept any complacency and laziness. We must always demand excellence and top-notch innovation.

Did you know that the other reactors at Chernobyl kept operating years after the accident? In fact, the facility is staffed, even today.

These two accidents have happened; they've caused a lot of economic damage. Of other damages, I'm not sure; I doubt that the scope of the damage is as big as is being portrayed by the doomers and gloomers. I sincerely doubt that Chernobyl, in the end, would have a maximum death toll of more than a thousand. Given the fact that there are a plethora of documentaries and videos on the life in Pripyat—the city closest to Chernobyl. The fact that wildlife flourishes, the fact that there are still people living there, the fact that you can watch YouTube videos of people who trudge around in the ruins and look for radioactive parts, all give me pause that it might not be as bad as advertised by zum beispiel Greenpeace or the Helen Caldicott's and Jim Green's of this world.

We have to take into account that we now have had three decades of hands-on experience with observing the effects of one of the greatest nuclear accidents in the history of our planet. We have also had the experience to observe what happens when a nuclear bomb gets detonated, just shy of 2.500 times. We should by now be quite confident about the repercussions of these events. I'm not saying that we should continue creating mushroom clouds and unwittingly destroying nuclear facilities, but we have the knowledge to be able to make well-informed decisions. We also know that the hysterical and magnified rhetoric that is being spewed by people that oppose nuclear has no basis in reality.

We know for instance that at Chernobyl probably only the direct responders died from the effects of extremely high doses of radiation, they were trying to stop a nuclear facility from burning down and at that moment the radiation near the facility was indeed very high. Many people do not know that Chernobyl had hardly any radiation shielding at all and containment facilities were absent. In the aftermath, however, no epidemiological evidence has been found to suggest that Chernobyl has contributed to any more radiation deaths except for a dozen or two, there are some hypotheses that predict a range of possible death tolls albeit with ranging degrees of certainty.

We also have to provide context here, contrast Chernobyl with the amount of deaths from smoking, cooking with sticks and dung, drinking contaminated water, inhaling ubiquitous coal-ash, consuming alcohol and we may only conclude that nuclear energy has an extremely limited negative influence on human health. Once we see the utter discrepancy between "nuclear" deaths and deaths from everyday exposure to non-nuclear —often self-inflicted— circumstances we have to acknowledge that to play on the fear of radiation is tantamount to needlessly terrorizing people. These people spread unsubstantiated nonsense and are thus fueling cognitive dissonance and irrationality.

One must consider that the nuclear accident at Fukushima has caused not a single radiation victim. We have yet to see the first radiation victims. The haphazard evacuations afterward, however, have claimed the lives of several hundred elderly that were pulled out of their well-heated and safe homes. Blind fear of radiation can and will kill you.

Science a la carte

According to Jim Green, an Australian anti-nuclear activist, there are *radiation deniers* out there. We have two ends of a spectrum here, people that have a rational look on the effects and scope of radiation exposure and those who think it is the direst threat on the planet. Jim Green belongs to the second category, and he has made it his job to point out which individuals belong to the first category.

"The nuclear lobby didn't even win the battle of the celebrities at COP21. James Hansen and other pro-nuclear celebrities put up a good fight against pro-renewable celebrities such as conservationist David Attenborough. But the pro-renewable celebrities raising their voice during COP21 included Pope Francis ... and he's infallible!"

Look at that! The fact that the *infallible* pope is pro-renewable doesn't mean anything to me, and it shouldn't mean anything to you, or to Jim Green. In fact, the pope is the leader of a pseudo-scientific death cult that worships beings whose existence cannot be tested because they have been put in the ethereal realm. Besides is this once again an argument from authority? Should we take the pope seriously because he is the pope? Or because he has been trained in the sciences? The answer to both questions is no. If he would present a solid well-evidenced case, we might consider it, but not because he has qualifications and a title. David Attenborough, on the other hand, is a slight disappointment to me, has old age addled his once so beautiful mind?

Jim Green continues in *"Pro-Nuclear Environmentalists and the Chernobyl Death Toll."* According to him, people like James Hansen, Mike Shellenberger, and Ben Heard are radiation deniers, which means that they, according to him, deny the possible greater scope of Chernobyl's disaster. The World Health Organization and UNSCEAR tell us that the immediate death toll from the accident is around fifty. This sounds ridiculously unbelievable according to Green. Green then wields the non-fatal thyroid cancer figure of

4000 and the expected non-fatal thyroid cancer figure of 14 000 and then tells you that he expects there to be much more deaths. This is an incredibly warped sense of reality; he cites non-fatal cases, adds another chunk of non-fatal expected cases and then conflates them with possible deaths. Is this a lie?

Consider this Jim Green quote:

*"Indeed, UNSCEAR itself co-authored a report which cites an estimate from an international expert group - based on collective dose figures and risk estimates - **of around 4,000 long-term cancer deaths among the people** who received the highest radiation doses from Chernobyl."*

And take a look at this excerpt from the same site he refers to in the same sentence:

*"About 4000 cases of thyroid cancer, mainly in children and adolescents at the time of the accident, have resulted from the accident's contamination and **at least nine children died of thyroid cancer**; however the survival rate among such cancer victims, judging from experience in Belarus, has been almost **99%**."*

So when Jim Green says that there have been 4000 long-term cancer deaths, is he actually lying? Is he dumb enough to provide a link to the correct figure?

Consider this excerpt from the same WHO page:

"Poverty, "lifestyle" diseases now rampant in the former Soviet Union and mental health problems pose a far greater threat to local communities than does radiation exposure.
Relocation proved a "deeply traumatic experience" for some 350,000 people moved out of the affected areas. Although 116 000 were moved from the most heavily impacted area immediately after

the accident, later relocations did little to reduce radiation exposure.
Persistent myths and misperceptions about the threat of radiation have resulted in "paralyzing fatalism" among residents of affected areas."

Yes, that is right; this is the resource that Jim Green chooses to use to drive home his own form of *paralyzing anti-nuclear fatalism.*

Also, consider that there are organizations that try to guestimate these figures by applying LNT-models. Where's the corroboration with reality? Where's the epidemiological evidence? Green loves to cite Greenpeace's 270 000 cancers and 93 000 deaths up until 2065.

Because organizations have been scrutinizing health records for decades now, and have been gathering epidemiological evidence and cannot corroborate additional deaths to the accident, it is justified to hold a skeptical position regarding the death toll from Chernobyl. And the same can be said for Fukushima, for which we have a radiation death toll of zero until thorough and unbiased scientific research proves otherwise.

This stings the anti-nuclear activists the most, the fact that the evidence to support their radiation-fear narrative is non-existent. And because they cannot believe it, they either start blaming others or construct some sort of ludicrous conspiracy theory.

— What is nuclear waste?

The brunt of nuclear waste is spent fuel, which isn't exactly an accurate name but if contemporary reactors are concerned, it is aptly named because you cannot get any more energy out of the fuel when you keep it in the reactor. The make-up of this spent fuel is about 93% Uranium, 1,3% Plutonium, 0,14% Actinides (Like Neptunium and Americium), and 5,15% fission products. The Uranium and Plutonium and the Actinides can actually be used to create more energy so about 95% of it can be re-used, in a sense making it a fuel rather than nuclear waste. All the long-lived waste that is normally considered a problem is actually a source of energy that can be tapped in certain designs of nuclear reactors.

Most spent fuel is simply stored in storage ponds or dry cask storage containers. There are also small quantities of Tritiated water and Xenon gas. These can both be stored in tanks and bottles until they decay. Concerns about Tritium are often raised, but if you consult these pages, you might want to reconsider if you think it to be as dangerous as the activists want you to believe it is.

Idaho State University—Tritium Information Section

United States Environmental Protection Agency—Radionuclide Basics: Tritium

United States Nuclear Regulatory Commission—Backgrounder on Tritium, Radiation Protection Limits, and Drinking Water Standards

Or this publication on the Scientific American Website: "*Is Radioactive Hydrogen in Drinking Water a Cancer Threat?*"

Tritium is used in everyday stuff like watches and emergency exit signs. It also occurs naturally and as such is dispersed throughout our food chain and water supply, which means that we ingest it regularly, from natural sources.

How much nuclear waste is there? And how does it stack up against other waste-streams? Over five decades of commercial nuclear power generation we've accumulated about 270 000 metric tons of spent fuel, contrast this to the 110 000 000 metric tons of toxic and radioactive coal ash that are produced annually in the US alone. Be reminded that coal-fired power plants release all of the pollution into nature through their smokestacks and other tailings that end up in coal ash heaps and coal ash ponds.

Nuclear waste, on the other hand, is mostly stored at the reactor facilities themselves, in dry casks or in spent fuel storage ponds. It just sits there; it is mostly inert and doesn't go anywhere. Besides spent fuel isn't non-fuel, it still contains masses of energy. Basically spent fuel and nuclear bombs can be turned into energy for peaceful purposes and this we should do! The incredible thing is that when (not if) we progress into the age of the Fast, the Breeder, and the Molten Salt reactors we will be producing masses of energy with incredibly marginal and manageable waste streams. Further chemical research will probably prove that there are applications for these *waste products* lurking in unknown territories.

The total world-wide nuclear waste could be stacked up and wouldn't be as big as an average football stadium, and I'm not talking about the big ones... It is that marginal. And only a small portion of it is the stuff that we should handle with great care, yet can manage quite easily. Volumetrically "nuclear waste" is like nothing compared to anything else in the world.

— Proliferation

Each time nuclear material gets moved, people go nuts. Consider for instance the Plutonium shipment from Japan to the US in 2016; this shipment immediately precipitated an anti-nuclear frenzy of epic proportions. It is the doom of our world, it is as if the devil has been caged and is in transit only to be freed by his evil minions.

Consider this simple fact, the world has been held hostage by nation states that have a great host of nuclear arms, they used to be tens of thousands. SALT and SORT have made sure that these numbers have diminished significantly. Nuclear weapons are being watched like a hawk, nations that intend to have these weapons are being watched like a hawk. Creating weapons-grade isotopes require special facilities and reactors. Commercial reactors do not create the volumes required nor are these isotopes easily separated from the spent fuel rods.

Therefore, the argument that creating nuclear energy equals proliferation is a non-sequitur.

The MSR is yet another boon against the proliferation argument, it simply turns these isotopes into energy, they remain in the salt, won't be separated and stay in there for long enough so that they themselves may fission. And with Transatomic's TAP reactor atomic weapons will be turned into energy, and if this process actually gets started, we may actually speak of the exact opposite, the reverse of proliferation—unproliferation? Deproliferation? Nothing is better at de-weaponizing than reactors that burn bomb-grade material. They eat it, they make it go away. And yes there will always be reactors that may produce amounts of isotopes that can be used to create weapons, but we have tons of possibilities to mitigate these issues and a lot of these are already in place.

— Terrorism

Have you an advanced missile system lying around in your basement? Is there a 120mm howitzer in your garage? Terrorism is one of the pet-peeves of the contra-nukes. We should be terribly afraid of terrorism. We should be afraid that a nuclear power plant would be a target for terrorists—which is a frequently seen argument. First of all consider the rigid and robust containment building that encases a nuclear reactor, consider the multiple layers between the containment building and the reactor. Consider the amount of strong explosives or highly advanced weapons required to breach these containment facilities, which are made of thick re-enforced concrete and steel. It would surprise me if a terrorist would manage to get any of these even near a nuclear facility, not to mention to deploy a device that cracks open a nuclear facility

136

and destroys the reactor core or one of its ancillary systems, which are redundantly available anyway.

Also, take note that newer designs like the Molten Salt Reactor are even more robust and safe. Crack one of those open and nearly nothing happens, barely any radiation would escape, and it would be soaked up by natural radiation levels anyway. Also, a molten salt reactor is encased in its own concrete pit and has a giant steel lid on top of it. Good luck getting through that.

— What about the theft of nuclear materials?

Any idea how difficult it is to create a bomb that would detonate and cause a nuclear chain reaction? Any idea what kind of facilities are required to build something like a nuclear bomb? Besides nuclear bombs are not results of commercial nuclear energy and cannot be bred from nuclear energy reactors without being noticed by the frequent and routine checkups by multiple global nuclear agencies. It is a chemically challenging process which is highly unlikely to be available to terrorists and the facilities capable of enriching uranium are in a state of constant surveillance and high scrutiny. And then there's this other element that can be used in nuclear bombs, Plutonium. Do you have any idea how hard it is to get nuclear grade Plutonium239? Minute concentrations of Plutonium239 exist in commercial reactor cores and spent-fuel but it is extremely difficult to extract it, in fact, it is encased within a multitude of other isotopes that do not permit you to utilize it to make a bomb, they are either neutron absorbers or isotopes that spontaneously fission, you would need to separate them chemically which is not done in a basement using chemicals you buy at the local supermarket or hardware store. I conclude that it is hardly possible to build a nuclear bomb through the theft of nuclear material from a commercial nuclear reactor facility; it simply is far too difficult and cumbersome.

Anticipating the apologia argument

One may argue that having radioactive materials alone is enough to create a *dirty bomb*. I would argue that you could put some C4 in a container filled with bananas and achieve the same effect from spreading minute traces of radioactive Potassium40. Or buy a million smoke detectors or emergency exit lights, or steal one-hundred x-ray machines. Sticking spent fuel or Uranium to a bomb doesn't do anything except spread fear. There's no need to be afraid of a dirty bomb—obviously do not stand within the blast range, which can still kill you...

Sadly, terrorists have now figured out that is far more effective to spread fear by killing lots of people with guns, as we have recently seen in Paris in 2015 and 2016.

— Correlation does not mean causation.

This is very important, and it's a logical fallacy to presume that X means Y, it's something that is very pervasive in the world of quackery and demagoguery. The fact that a lightning bolt struck a tree doesn't mean it is infested with evil demons yet the impressionable and the scared and particularly the superstitious could easily be convinced that it was.

So it is also with radiation. Demagogues want you to believe that that it is nothing but evil and dangerous, but is it? The first thing we need to learn about radiation in high-school is that it is everywhere. It is in the sun, in the planets, in the comets, in space dust, in space, it is in the trees and in the soil, in our food and in our bones. How did it even get there? The periodic table of elements is just one layer of many, did you know that there's layer upon layer of different atoms of the same kind (elements) yet with minute differences? We call these isotopes. Potassium, for instance, is an element that is common in this world, we need it, it is a necessary part of our diet, it also exists in the earth's core where minute percentages—yet a huge volume—of the radioactive isotope of Potassium decay and in the process lose some of the

138

energy that kept them together in the form of heat. It is one of the isotopes that keep the core of the Earth hot. Traces of this potassium isotope—potassium40—also exist in our food, bananas for instances are particularly rich in potassium. Does this mean that bananas are bad because they are radioactive? No, bananas are healthy.

What about other sources of radiation? Thorium and Uranium are quite common in the earth's crust, these elements are everywhere and quite common in certain rock formations such as granite and marble, the wondrous materials we use to create all these magnificent buildings and floors and kitchen tops. What about clay? Some pots are radioactive, particularly the ones that have a deep red color finish. What about Radon? It simply gets exhaled into the atmosphere by the Earth, and we breathe it. What about cosmic radiation? Did you know that light is radiation and that it consists of many different kinds and wavelengths and particles? Did you know that pilots get more radiation exposure than people that never fly? The atmosphere absorbs portions of this cosmic radiation. Astronauts get the most breathtaking views of the stars and the moon and the Earth all while being bathed in radiation levels some may deem deadly. Is there any evidence to suggest that people who left our atmosphere have extraordinary cancer levels? We are slowly arriving at the point where we may conclude that exposure to radiation does not equal a high chance of getting cancer or dying.

The linear no safe threshold hypothesis is one of these ideas that edges very close to trying to prove any amount of radiation exposure causes cancer and it is the stick that many of these demagogues like to beat nuclear energy with. Consider this simple yet effective analogy, LNT suggests that falling on your butt a thousand times is as dangerous as jumping off a ten story building while trying to land on your butt. It is a total non-sequitur, it does not follow logically that no dose of radiation exposure is safe or beneficial and on the other hand it also does not follow that any dose of radiation exposure is safe or beneficial. This is something that needs context and is a very complex and mainly biological issue.

So what is it then? It seems that too much of a good thing can kill you.

There are counter hypotheses such as Hormesis—the fact that damaged cells start repairing themselves. Or consider the pervasive use of radiation in the medical world to diagnose illnesses, and body functions to cure patients. How can the LNT-hypothesis be true if radiation used by doctors keeps people from dying? If the LNT-hypothesis were true, all doctors would be liable for using any radiation to diagnose and possibly cure ailments.

— How does this stack up against nuclear power plants?

First, consider that nuclear power plants are highly regulated and expected to keep the radiation from its fission processes contained. Also, contrast this with the fact that Coal-Fired power plants are not held to the same standards and emit considerable amounts of coal-ash that contains traces of uranium, thorium, and radon gas. Here we have a double standard that shows a contradiction, if radiation really is the issue, then why are coal-fired power plants not held to the same standards as nuclear power plants? If the LNT-hypothesis were real we could take the government, the regulatory agencies and the fossil fuel companies to court and sue them into oblivion, but that's the point isn't it, we don't do it. Why? Because science tells us that radiation is everywhere.

If the LNT-hypothesis was true the epidemiological evidence for vastly increased cancer rates in Chernobyl, and Fukushima should be overwhelming, instead they are not. It is true that elevated rates of non-fatal thyroid cancer have been observed, but they are non-fatal. Besides they are not of the magnitude that warrants such mass delusion and fear mongering. Context is always required. Let's contrast the cancers that have possibly been caused by the Chernobyl accident with annual death tolls from other causes.

Chernobyl   : confirmed 50 dead, 4000 *non-fatal* cancers

The following figures are annual worldwide death-tolls

Smoking   : 6 Million of which 600 000 non-smokers (WHO)
Alcohol   : 3.3 Million in 2012 (WHO)
Car-accidents   : 1.25 Million in 2013 (WHO)
War   : 180 000 in 2014 (IISS)
Malnutrition   : 3.1 Million children per year (WFP)
Malaria   : 438 000 deaths (WHO)
diarrhea   : 1.5 Million (WHO)

Do you think that nuclear is as bad as some of these people advertise?

Let's venture into environmentalist mystical hypothetical-land: suppose Chernobyl caused 5000 deaths per year since 1986, it would have claimed 150 000 deaths by now—which terribly exaggerated. Let's extrapolate all the other figures over the same timeframe backward over 30 years.

Smoking   : 180 000 000 vs. 150 000 = 0,083%
Alcohol   : 99 000 000 vs. 150 000 = 0,152%
Car-accidents   : 37 500 000 vs. 150 000 = 0,400%
Malnutrition   : 93 000 000 vs. 150 000 = 0,161%
Malaria   : 13 140 000 vs. 150 000 = 1,142%
Diarrhea   : 45 000 000 vs. 150 000 = 0,333%

See how ridiculous these juxtapositions look? And I've even made them worse than any nuclear-doomer would dare do.

Anticipating the apologia argument

There will be some that will play the argument from authority card here, and Jim Green is one of them and has already done so. Pay no heed to their vain and sanctimonious cries, keep a clear head, remain reasonable and go and look for the evidence that has been presented by UNSCEAR and the WHO. Don't take my word for it, because I am but a fool who tries to understand the world, I merely share with you the steps I take during this quest for more knowledge and understanding.

Trying to make the case for shutting down the entire nuclear industry, like Jacobson does in his feasibility study, is the stupidest thing we could be doing right now. Not only would we be severely crippling our potential to defeat fossil fuels, we would practically shut down a great deal of the medical world, and we would also cause severe stagnation to any future developments that could enhance and improve all issues that *they* think unfixable. Consider one of these developments to be the assembly-line-like MSR production proposed by Thorcon and Terrestrial Energy. Or consider the WAMSR/TAP reactor designed by Transatomic Power that can rid us of practically all spent fuel i.e. nuclear waste, and nuclear bombs, or the LFTR that will bring Thorium breeding to the table.

If we are honest with each other and acknowledge that spent-fuel still has a lot of energy, why would we let it go to waste, or think of it as a problem? This is what the antagonists do. They like to raise the problem of waste while ignoring the possibilities it presents to us. It's yet another nuclear-energy-antagonism-paradox, make a problem out of it but not accept any possible solution.

It is time that we emancipate ourselves from this foolish and dangerous anti-nuclear ideology. Not only is it unfounded and fraudulent nonsense, but it is also dangerous nonsense that keeps us from addressing the true issues at hand. Instead of seeing nuclear energy as the solution that it really is they create this additional specter and—to add insult to injury—fight nuclear power rather than to present credible solutions to anthropogenic climate change and ocean change and the looming arctic methane apocalypse and eventually a possible mass extinction event.

In writing this chapter and naming it "anticipating the apologia argument" I realized that nuclear energy actually has nothing to apologize for. The good of nuclear energy dwarfs the bad by orders of magnitude comparable to the *juxtaposition of an Elephant and a Mouse*, and we should continue and improve and move forwards, rather than regress as some of these people wish us to.

The power of the force

The most elemental forces of nature and the propensity to create new elements through fusion have led to the creation of the elements we know today. These elements have been adopted into the beautiful Periodic Table—Which is the culmination of the works of Antoine-Laurent de Lavoisier, John Newlands, and Dmitri Mendeleev who eventually published it in 1869.

During the process of Primordial Nucleosynthesis and the continuation of it during the ensuing eons of time, almost all elements available in the solar system have been created in the first and most magnificent stars. Among these elements, we find the radioactive isotopes of Potassium, Uranium, and Thorium. These elements are ubiquitous in the universe and present all around on our own planet and as such we are constantly subject to radioactivity. Radiation also comes from the sun and from space itself, the culmination of all the nuclear forces working seemingly and unendingly all around us.

Our universe is held together by the four fundamental forces of nature: The strong nuclear force, the weak force, gravity and the electromagnetic force. Which force is most pertinent to our current endeavor? The strong nuclear force. It keeps the nucleus of an atom together—the protons and the neutrons—and this force is the key part in unleashing energy from the nucleus, the source of the energy we get during nuclear fission and fusion. Simply consider the "basic fact" that protons don't want to be next to each other. As soon as you disrupt the equilibrium in the atom it will split and it will break apart into fission products and neutrons and heat, and this is the little secret behind the production of nuclear energy.

We all know that we can build bombs using some of this "stuff".

The detonation force of these contraptions has multiple effects on the human psyche, at first it is humbling and incredibly terrifying, but if you think it through from a more rational perspective you learn that through the utilization of very small quantities of potent elements we can tap into tremendously large amounts of energy. We can build these bombs using certain radioactive isotopes which are very hard to get, you have to separate them from the other naturally occurring isotopes, and if you want a very strong bomb, you need to breed them. You cannot dig up weapons-grade plutonium anywhere on the planet. It only comes out of specialized nuclear reactors that depend on a very specific type of Uranium that needs to be refined and enriched in the first place.

Do commercial nuclear reactors create bomb-grade radioactive isotopes? Yes and No. Yes, there are minute quantities of isotopes that can be used in nuclear bombs, extracting it is not easy, or cheap and does not get done, nor does it go by unseen. Nuclear agencies would know if a reactor would be used to create weapons-grade materials for proliferation purposes. Expensive chemical facilities are required to refine these isotopes and obtain them in such quantities that you can use for proliferation purposes. The "*nuclear energy equals nuclear weapons*" argument is a total non-sequitur that requires a ton of qualifications and additions to even remotely become true.

Let's assume that we are never going to use these weapons ever again. What will we do with them? Store them for an eternity? Dismantle them? Suppose we chose to dismantle them, can we do something with the fissile material that is present in these bombs? Why don't we stick it in special reactors—like Transatomic's

The power of the force

WAMSR/TAP reactor—and make energy out of it? The science is there, and we have people who want to (and can) make it happen. We could literally reduce our stockpile of nuclear weapons by turning these bombs into energy, why aren't we doing it? The answer is, of course, obvious, but I'll leave this line of reasoning here simply to make you hungry for this solution. We can do this and should, and with the implementation of this technology could quite possibly remove a lot of anxiety in the world and educate a ton of people.

Nuclear energy generation started in the 1940's and took off with the conception of the Nautilus. From the inception of the submarine reactor, the fate of nuclear energy was sealed for a long time. The reactors that are being used in submarines and big aircraft carriers are very well constructed and serve their purpose well; they produce reliable and safe energy that can sustain crews under water for weeks. Also, consider that there are half a dozen Russian icebreakers that are being powered by small nuclear reactors.

In the decades to come commercialized versions of the ship reactors have started to power parts of civilization. Most of these are Pressurized Water Reactors which have been around since the 1960's. We consider these to be Generation II reactors. The generational number is an indicator of how modern a reactor is. Generation I reactors are the first, the prototypes; Generation II are the first reactors designed for commercial use; Generation III are logical iterations of generation II designs, they are evolutionary, there's also Generation III+ which is reserved for designs that include improved passive safety features and are completely standardized.

I'm not really interested in discussing Generation II reactors since we're now slowly and gradually moving away from them. Generation III and III+ designs will remain dominant until several generation IV designs will prove to be economically more

146

interesting i.e. cheaper than coal and everything else, then they (generation IV) will sweep the market, and the face of nuclear energy will forever be changed, giving us a far more optimistic outlook on the future.

If you want to know more about the different generations of nuclear reactors you are best served by reading the Wikipedia page on Generation IV reactors which have an excellent graph that shows a timeline and a couple of designs. You can find a ton of links in the appendix of this book, if you're done reading them, you should be completely up to speed.

In the meanwhile a company called Lightbridge has developed a new solid-fuel principle that will stretch fuel-efficiency for solid fueled reactors and we may expect them to get a kick of efficiency and profitability. Be sure to check them out, very interesting stuff.

The true revolution, however, can be found within the realm of Generation IV designs, one of which is my personal favorite: the brainchild of Alvin Weinberg, the Molten Salt Reactor (MSR). The MSR is actually quite an old design; it has been around since the nineteen-sixties. A very challenging question spurred Weinberg to start thinking about a compact and lightweight reactor design that didn't require a separate cooling loop, and that could be used to power a bomber without the hassle of solid fuels. The result was a reactor in which the fuel was dissolved into a molten salt, which was simultaneously the working fluid and the coolant. Arbitrarily a molten salt has a melting point somewhere around 400 degrees Celsius, which means that if the temperature drops below 400 degrees Celsius the salt solidifies, and this is an excellent characteristic for a fluid that is both the working fluid and the coolant at the same time. This means that if there's a loss of power the fluid can be drained into a special drain tank, which has no moderator and this means that the fission process in the fluid will stop and the fluid itself will start cooling down until it solidifies and stays that way until someone heats it up again. This process completely removes the fear for meltdowns because there is no more solid fuel that can melt. This makes a sudden shutdown of the reactor a rather safe situation in contrast to the fleet of Generation II light water reactors that are in operation today and require constant and active cooling processes to prevent core temperature from rising uncontrollably.

What is fundamentally different about an MSR over current predominantly light / pressurized water reactors?

LWR/PWR fuel is solid, which means that fission products get trapped in the fuel and prevent further chain-reactions, and this means that the fuel efficiency is low, and "waste production" is high, yet this waste is untapped nuclear fuel, a nuance that has to be made.

MSR fuel is a liquid, the Uranium is dissolved in a salt, this means that it can move around freely, and fission products can be extracted and because the Uranium is dissolved the fuel-efficiency improves dramatically from a 3~5% in an LWR/PWR to roughly 95% in an MSR. Through these efficiency gains, we may finally realize a sustainable form of nuclear energy that will be functional to human civilization not for a century but for millennia to come.

The MSR provides fuel-versatility, it is possible to run MSRs on different salts, different Uranium isotopes (not just the rare U235), Fertile material like Thorium232 can be used in a breeding cycle —which has multiple benefits like the possible extraction of Bismuth213.

An LWR/PWR needs liquid cooling, and in order to keep the water a liquid it is pressurized, this is to keep it from becoming a gas, which it normally does at 100 degrees Celsius. High pressure means that if there's a rupture anywhere, the pressurized and superheated water that escapes the loop will gasify immediately and this causes its volume to increase exponentially. Therefore, an LWR/PWR has a special containment building that accounts for these possible — yet very unlikely— events. The MSR, on the other hand, works at atmospheric pressure which means that if there's a rupture in the fuel loop there will be no blow-out, the fuel itself will solidify once it exits the loop. This means that there's no requirement for a containment building that has to cope with an explosive pressure release from the reactor system.

If you are afraid of meltdowns and want them to go away, the MSR is your answer, it simply cannot meltdown, a catastrophic failure would lead to the opposite, a freeze up and this is precisely what you want. The MSR loop has a freeze-plug, which is also a salt that remains solid as long as it is cooled. Once the power fails and thus the cooling, the plug melts and all the salt from the reactor is drained into a tank where the chain-reaction stops, the salt cools down in a matter of hours and turns into a solid that is inert. The only heat that is left will be from the decay of fission products, and this is not enough to melt the salt again because this heat is removed from the drain tank through natural convection.

Finally, the MSR is easier to mass-produce, which is the final and most valuable asset in its toolbox. Humanity is going to need a lot of these if we want to end our dependence on coal and other fossil resources.

We now know that the MSR has a lot of promise, let's take a look at some of the promising designs that are out there.

The LFTR, the first true Thorium design by FliBe Energy.

FliBe Energy is run by Kirk Sorensen, arguably one of the most intellectually talented engineers I've ever seen talking about technology. I've met Sorensen once, we didn't get much further than a handshake, and my attempts to converse with him over e-mail were unsuccessful, but given the fact that his trying to usher in a revolution in nuclear energy, I can see past it, but I sure hope to get another chance...

Kirk is working on the LFTR, which is an acronym for Liquid Fluoride Thorium Reactor—which is probably the most advanced MSR design to date. The LFTR includes chemical processing, which is required in a two-fluid breeder reactor. What it does, in a nutshell, is this: it turns Thorium232 into Protactinium233, which decays into Uranium233 which then can fission so it may release some of its energy content and this energy will be used to generate electricity. Chemical processing maintains the working and blanket fluids which ensure optimum and continuous performance.

FliBe, in the long run, envisions modular building principles, both in reactor modules, chemical processing modules, multiple module facilities, etc. This means ease of production, ease of installation, ease of scalability, and ease of swapping.

One of the unique opportunities of the Thorium fuel cycle is a byproduct called Bismuth213 which can be used to treat certain forms of cancer. Bismuth213 only exists in the decay chain of Uranium233, which only exists if you breed Thorium232. And this could be a tremendous boon for medical research, and by not developing reactors to run on the Thorium fuel cycle we postpone a development that could potentially save millions over many years to come.

Gordon Mcdowell has been on Kirk's tail for a while now, he has done an excellent job in capturing this technological adventure and helping us understand what is required to get where we want to get. Firstly you absolutely should visit www.flibe-energy.com, it is expertly written, it is understandable and gives you a better in-depth view on the technology than I could ever give you, secondly type in Kirk Sorensen or LFTR in the YouTube search bar and you will get a tremendous amount of videos in which you will become enriched with details about the history of nuclear energy and its bright future.

I fully support Kirk Sorensen's view that the Thorium cycle will provide in the long run. We're pushing for the Thorium-age, and I hope he manages to get this vision realized and as such the culmination of Alvin Weinberg's intellectual ideas of a nuclear energy future that brings prosperity and stability for all.

What is so exciting about Thorium?

There are a couple of isotopes that we can use to create nuclear energy in a reactor: Uranium, Plutonium, and Thorium. Plutonium almost doesn't exist in nature, so that one's out. We use the extremely rare stuff to generate nuclear energy, the Uranium-235 Isotope which is less than a percent of all the Uranium on Earth. The rest is Uranium238 which cannot be used in contemporary nuclear reactors. The reserves of Uranium235 are only good for a couple of decades if we keep producing energy with it as we've done the past. When we can switch up to Uranium238 we extend this with hundreds if not thousands of years. Thorium is another element that may be used in a breeding cycle, which means that you turn Thorium232 into Uranium233 by letting it absorb a neutron and turning it into Protactinium233 which decays into Uranium233 which may fission if hit with another neutron. The beauty of this process is that it runs at high fuel-efficiency. But even more exciting is the fact that we already have tons of it sitting about, and therefore, no mining is required for a very long time. Thorium is far more abundant than Uranium and this means that we may stretch the nuclear fuel availability for tens of thousands of years, maybe even millions of years, well enough time to go seek for these elements in space and to get nuclear fusion going. How's that for sustainability? Simply visit www.daretothink.org and look for this article: "How long will our supplies of uranium and thorium last?" Dare to think dot org is run by my fellow countryman Gijs Zwartsenberg.

Normally you would have to burn millions worth of fossil fuels to achieve the same energy generation as you can get from pennies worth of Thorium. It is that abundant and cheap. Why? Well, it is considered a waste product from mining activities that causes a lot of headaches to those who have to dispose of it. People literally don't want it!

Thorcon—The mass produced MSR(e)

Over the years that I've become interested in nuclear energy I've become enamored with Thorcon Power and its very down-to-earth approach to get Molten Salt Reactors build. They want it now, and they work hard to get there. And that's why they aim for maximum flexibility and modularity in their design. Where some startups look for optimum scenario's straight out of the gate, Thorcon is willing to start on different principles, like a different fuel-salt (NaBe - Sodium Beryllium) and a different fuel mixture, and as time and innovation and availability progress, they can switch to the optimum fuel-salt (FliBe - Fluoride Lithium Beryllium) and different fuels, ultimately thorium.

I've come to know Lars Jorgensen and Robert Hargraves—who are both involved with Thorcon Power—by contributing to a mailgroup we're in and I recommend taking note of their thoughts on nuclear energy and how to make it cheaper than coal—something which Robert has explained concisely in his book: "*THORIUM: energy cheaper than coal.*" Lars can be seen on multiple YouTube videos, a particularly good one is: "*Thorcon: A Thorium Molten Salt Reactor System that can be built now - by Lars Jorgensen @ TEAC7*"

Most reactor designs nowadays are still quite intricate and will be constructed on site completely which means that they are quite labor intensive. If you look at equally complex technologies such as airplanes and ships and automobiles you see that they are being built in a modular fashion. This means that there's an assembly line for the wings, an assembly line for the hull, an assembly line for the engines, an assembly line for the fuselage, etc. etc. All these components get built more quickly and come together effortlessly which greatly increases deployment speed. Where building a car by hand with 10 very skilled people may take a month or two, the same car only takes a day or two to get built on an assembly line.

This is what Thorcon envisions for their nuclear reactor facilities. They compare their process to ship and airplane building. They envision a shipyard-like production facility at which all the individual components of the reactor facility get produced, tested and shipped. Their units can then be loaded onto barges which subsequently move the modules to the sites where the nuclear power plants are being constructed.

The nuclear facilities of the Thorcon design are subterranean / under the ground, which makes building them quite easy. First of all, you may *dig a hole*, construct the concrete bucket in which the entire facility gets built and then install the modules that have been constructed at the production facility by lowering them down into the facility and hook them up.

One of the great benefits of this is that construction times of reactor facilities get reduced, but also that components do not need to be repaired on-site, but will be swapped on-site. This also enhances efficiency.

Remember that I talked about the possible deployment speed of future nuclear power plant designs? Consider this excerpt from the website of ThorCon Power.

"The entire ThorCon plant including the building is manufactured in blocks on a shipyard-like assembly line. These 150 to 500 ton,

fully outfitted, pre-tested blocks are barged to the site. A 1 GWe ThorCon will require less than 200 blocks. Site work is limited to excavation and erecting the blocks. This produces order of magnitude improvements in productivity, quality control, and build time. ThorCon is much more than a power plant; it is a system for building power plants. A single large reactor yard can turn out one hundred 1 GWe Thorcon reactors per year."

Thorcon power has an agreement with the government of Indonesia which brings them closer to realizing their vision of the revival of the MSR[e] which was developed by Alvin Weinberg.

Note the one hundred figure, this makes the targets we've stated earlier come within reach.

Transatomic—The waste burning MSR

When I first saw Leslie Dewan and Mark Massie talking about their reactor, I was astounded, two youngsters advocating a new kind of reactor that could *eat* nuclear waste and nuclear weapons. They had an excellent analogy about our current fuel-cycle, I am not quoting verbatim, I'm paraphrasing:

"Suppose you're a hungry student and you make yourself a nice sandwich and you take one bite, and suppose you are still hungry but instead of eating the rest of your sandwich, you make another one and take yet another bite and so on and so on."

This is a very easy to understand and accurate analogy on how we *consume* nuclear fuel. The contemporary solid-fuel cycle is only 3~5% efficient, which means that we do not use the other 95% of the fuel, and it is regarded as spent fuel or nuclear waste. When I learned this, and I didn't know this before, I suddenly became very hopeful about nuclear power and it kindled an interest within me and made me start to read about it, up until the knowledge I've gathered so far.

Science a la carte

At first, their design was known as the WAMSR, an acronym for waste annihilating molten salt reactor, but now they market it as the TAP reactor. When you visit their website you can see a dashboard that shows you their progress. Given the fact that their reactor is specifically designed to deal with spent fuel and weapons-grade materials makes it an excellent proposition, there's no mining required to get any feedstock, in fact, it is estimated that all the spent fuel that can be used in TAP reactors made by Transatomic contains enough energy to power the entire civilization—taking into account population growth and increasing demand—for seven decades. This is the kind of technology that can be used to wash away all the ills that the opponents of nuclear energy ever so often pose as insurmountable.

Terrestrial Energy—The IMSR

Last but most certainly not least in this list of North American MSR startups is Terrestrial Energy, which is tied on pole-position with Thorcon. Terrestrial also has its eyes on modular designs that can be produced on a large scale. They envision complete reactor units that can be transported by road. Their design is slightly different from the other designs insofar that it doesn't dump its salt into a drain tank when the power fails, but it simply keeps it in the reactor unit, where the heat will be absorbed by a blanket of salt. By using this method the reactor can be restarted quite easily.

Terrestrial is a Canadian company and as such can play in a different ball field than its American counterparts and this brings about several advantages. For one, the Canadian regulatory framework isn't restrictive or single-minded, in fact, they let you make your case and they will examine it and from there on you can work forwards. Terrestrial is now in the game, they have gotten grants from the government to perform research, and they have acquired strong partners such as Caterpillar, Energy Northwest, Ontario Power Generation, PSEG, and the Southern Nuclear

The power of the force

Operating Company. Also, they are gathering intelligent and talented people from all over the world like fellow dutchman Teun Rodenburg as lead scientist, and the Australian Ben Heard as environmental advisor. Ben is the man who gave me the final push into writing when my personal outlook was dim and gloomy.

If you want to see the impact Terrestrial is having simply consider this document by Stephen Tindale, Katherine Chapman, and Suzanna Hinson called "*Next steps for nuclear innovation in the UK*" which you can find on the website of the Weinberg Foundation (http://www.the-weinberg-foundation.org).

At this moment, it is unsure when Terrestrial will finally commercialize and mass produce their reactor, but I am confident that it will happen within ten years or so, and that will be just in time to make a significant difference.

I highly recommend searching for Dave Leblanc on YouTube; you will find a great host of excellent videos in which he explains the principles behind Terrestrial's design and how far they have come.

We are nearly finished looking at MSRs, there are still quite some startups left, one of which is Copenhagen Atomics from Denmark and another is Aristos Power from France. This last startup is run by a good friend of mine but let's have a brief look at Copenhagen Atomics. They have a very interesting design philosophy: they envision a modular design that consists of adapted shipping containers that when coupled together form a nuclear power plant. These containers will be lowered into a concrete pit where they will be hooked up together in order to form a complete nuclear power unit. Once again we see a modular design that facilitates ease of manufacture, installation, and exchangeability. And think about how awesome this idea is, it is the ultimate form of integrating modularity within an already existing transportation system because all these modules are built in standardized shipping containers. Copenhagen Atomics has made a system that can be deployed easily, can be deployed on a bigger scale, and this makes their idea very promising. They envision building these on existing nuclear facilities, and when demand rises add units accordingly. It's a very interesting company and worth the while looking at. Google Thomas Jam Pedersen and watch this YouTube video: "*Making safe nuclear power from Thorium | Thomas Jam Pedersen | TEDxCopenhagen.*"

Why mention Aristos Power? Not just because of my personal connection but because they are at the very beginning of starting up and it is exciting to watch them make the necessary steps to become a full-blown contender in this field of expertise. Let's see what they are up to:

"*Our reactors are compatible with all other energy sources both base load and intermittent as they are designed to be capable of providing baseload power as well as load following and even peak power for short intervals. The current designs being developed by our company are of the TTR type (thorium thermal reactor). The fuel is LEU (Low Enriched Uranium) plus Thorium in a molten salt environment. Our lineup of reactors: The TTR mark 3 MSR: Type 10 at 10,5 MWe ; Type 20 at 21 MWe ; Type 30 at 31,5 MWe. The TTR mark 3 cores are designed for small communities, powering smart mini-grids or isolated locations where the cost of expanding the grid is prohibitive.*

The capital cost of the TTR mark 3 makes it cost competitive with NGCC power generators as well as small wind farms and solar farms, with the added benefit of being a baseload, dispatchable power source which can be adapted to the needs of each community. The TTR mark 2 MSR This is a grid power reactor which comes in only one capacity form, 560 MWe, It is the ideal solution for grid operators, medium sized cities, and medium-sized industry hubs. This design will be capable of producing electricity for as low as 2 cents per kWh, making it competitive with coal, hydropower, and legacy nuclear power plants (LWR tech). It can be used as well as a source of process heat.

The TTR mark 1 MSR This 2200 MWe power reactor is our flagship, It is specially designed to incorporate a number of innovations which will make it highly flexible and competitive for the foreseeable future to other Gen 4 reactors being developed, while still retaining the capacity for further upgrades. This is the ideal power reactor to power up large urban agglomerations and large industrial hubs.

We are in the process of doing the neutronic modeling of the mark 3 and mark 2 cores. In the near future we will try to co-opt a university/laboratory where material testing might begin, the next step is building a mark 3 core type 10 which will serve as a technological demonstrator. Initially, it will not be loaded with fuel, it will be a heated core to simulate thermal transfer in real life conditions. At the same time, we will start the licensing process.

The next stage will be replacing the heating core with a core loaded with nuclear fuel. this core will be tested for 2 years, afterwards it will be defueled, extracted from the reactor (which will also serve as a simulation to real life core replacement), examined to evaluate possible core damage and reinserted and run until an eventual core damage occurs, this will enable to evaluate the maximum possible lifetime of the core, while also simulating for worst case scenario core breach and reactor shutdown, as well as cleanup operations.

If all goes well the next step is to deploy to the market the TTR mark 3 type 10 and its larger versions the type 20 and type 30. After finalizing the RD & D of the mark 3, work will start on the first Mark 2 core, with the same stages to be followed as the mark 3. This will become the workhorse of our product line. There are currently no plans for the near future to bring to market the mark 1 core, unless market interest will be sufficient to warrant this development. Our design utilizes a very cheap molten salt, and the same goes with other reactor materials, which will make our concept be a factory built, easy to build, cheap to build, and cheap to deploy design.

The core of the mark 3 will be designed to be replaced every 2 to 4 years, operation which will be done on a timescale shorter than the refueling of legacy nuclear power plants (LWRs), thus reducing the downtime of the power plant and strengthening the economic case for our design. The TTR designs brings to the market a number of innovations unseen until now in other designs, innovations that greatly reducing costs and at the same time greatly improving safety and ease of operation. Aristos power is committed to make nuclear not just a sustainable source of energy, which nuclear power already is, but also to revolutionize the sector by making its reactors the first of a kind renewable nuclear reactor."

An increasing production of MSRs will supply many opportunities for renewed discoveries. Consider leaps in material sciences and chemistry. For instance, Hastelloy-N has been developed by Oak-Ridge National Laboratory specifically to serve as a material that is suited to build Molten Salt Reactors. Or simply Google Dr. Stephen Boyd or look for his videos on YouTube, which can also be found on Gordon Mcdowell's Thorium channel. Once chemists, material researchers, and engineers get to be integral to the mass production process of MSRs the learning curve will steepen, and new discoveries will be more forthcoming. It's the benefit of hands-on experience and increased research and development.

What does nuclear have to offer to the medical world? Medicine benefits hugely from radioactive isotopes, isotopes that are only available thanks to specialized high-flux reactors and a few commercial reactors.

First, take a look at these websites and take note of the sheer volume of radioisotope goodness in the world, goodness bred from nuclear reactors, which without the nuclear industry wouldn't be possible at all.

Note: Why would I ask you to read these sites? Because they convey the message far better than I could, and I want to keep the page count down, this was supposed to be a short read...

www.world-nuclear.org — Radioisotopes in Medicine
http://www.cleannuclearpowersafehospitals.com/
http://www.radiologyinfo.org — General Nuclear Medicine
Wikipedia—Nuclear Medicine

Without radioisotopes, we wouldn't have had X-ray machines, MRI's, or CAT-scans. We wouldn't be able to do certain organ function scans. We wouldn't be able to non-invasively diagnose cancer (oh the irony). The capability of certain vital or palliative treatments would be lost. Millions of people worldwide benefit hugely from the availability of nuclear medicine.

Bruce Power runs the world's largest CANDU nuclear facility with eight reactors. This facility has an interesting radioactive *by-product* that is used for sterilization of hospital equipment, food, and mail: Cobalt-60. The strong gamma rays emitted by Cobalt-60 can also be used in pest-control which might prove helpful in the plight against the Zika virus which is plaguing parts of South and Middle-America.

The principles of the nuclear force help us generate electricity, thermal energy and a great host of useful isotopes that are essential. As such nuclear energy is a great force for good in the world. Several reactor designs may propel humanity into a future of plenty and stability and progress. It starts with bridging designs like the AP1000, the EPR, the APR1400 and then quickly move on to MSRs like the Thorcon Reactor, the IMSR, the TAP Reactor, and the LFTR. We may also expect success from Bill Gates's endeavor with the design that Terrapower will test in China within the coming years. And take note of other startups such as Copenhagen Atomics and Aristos Power, and many more.

Urgency

If we keep burning fossil fuels and keep disrupting the hydrological and the carbon cycle, bad things are bound to happen. The sheer volume of scientists that have a firm grasp on what is going based on solid scientific evidence should give policy-makers and law-makers enough reason and mandate to help us transition towards a more stable future; one of plenty and stability; one in which every human being gets the chance to maximize its potential. Energy poverty simply will not cut it and yet this is what some are trying to advocate. Helen Caldicott is one of these people, she rambles on about useless star trek doors a-swooshing and refrigerators a-humming, and her love for Mozart, who wrote his excellent and mesmerizing music by candlelight, as if that would be a perfectly fine future for all of us.

As we speak, a tug-of-war is going on between people of a more rational persuasion and those who believe things to be possible by sheer willpower. This book tried to give you a more detailed view of this ongoing discussion. This tug of war is going to be won by rationalists in the end, I'm quite sure of it. I have to admit that I have become biased in favor of nuclear energy over renewable energy, but I think I have made perfectly clear how I arrived at this point. It is not a mere matter of blind faith, but precisely a lack thereof, but also, on the other hand, an unquenchable thirst for more information and understanding.

We have upset the carbon- and the hydrological cycle and these cycles are fundamental to life on Earth. Through these disruptions, we might cause a new mass-extinction event that could eventually lead up to the demise of civilization as we know it. In order to keep this from happening, mankind has to change course within a very short timeframe. Even though our leaders organize climate-change summits, it is not evident that there's going to be any noticeable change in the way we consume energy. In fact, if we look at our current state of energy affairs one may only conclude that it is the demand that has an influence on the carbon emissions and not the implementation of non-carbon emitting technologies. How can it be that despite the additions of countless of windmills and solar panels the furnaces still keep burning lignite, coal, biomass and gas in order to power modern civilization? It is because of the minute impact these renewable technologies have. Wishful thinking does not provide us with a reality that is in tune with our desires. In fact, a scientific knowledge and understanding through gathered evidence is required and only by adhering to these principles can we make well-substantiated and reasonable decisions that are in tune with reality and will help us avert possible catastrophes.

Consider this *fact*, despite the attempts of some to envision a declining demand through the implementation of some tricks in accounting and caps, it is predicted that energy demand is going to rise. The case for a rising demand is easily made. About two or three billion people live in circumstances of energy poverty, unavailability of potable water, unavailability of good sanitation, little or no medical care, poor nutrition, and subsistence. Fertility rates amongst these people are highest, which means that the amount of people living in troublesome circumstances is set to grow. Also, consider the fact that fresh water supplies are dwindling and despite all that economies want to keep growing. The case for a growing energy demand is much stronger than the case for a declining energy demand. And even if all these poverty stricken people would slide in with higher efficiencies, demand would still be rising, regardless.

Urgency

How are we going to fill the gap of this growing demand? I suppose it is reasonable to assume that energy demand is going to rise somewhere between 50 000 and 100 000 TWh. Suppose energy demand by the 2050's is 250 000 TWh pure electric—which would be one of the most optimal cases albeit practically impossible because there are some thermal processes that are very hard to do on electricity alone—and suppose that we have to choose a mix of technologies that ensures that we do not rip open all the lands in order to get the materials needed to build this energy mix.

Since we've put a tremendous amount of carbon in the atmosphere we have to deal with a couple of problems at once: First, how are we going to lower emissions drastically? Second, which energy sources pack enough punch to help us gain enough energy generation to fulfill that demand of say 250 000 TWh per year? Third, how fast can we deploy these technologies? Also, let's not forget that we may have to desalinate water and capture and sequester carbon as well, which is going to drive the demand upwards even more, perhaps even significantly.

The third question might be the most important one to answer first because that will eventually determine whether we will be successful or not. Although many of those pro-renewable people would like you to believe that money is the deciding factor, this is a non-sequitur. Deployment speed and the imposition of natural restrictions on raw materials will determine whether we can reasonably expect any scenario to become successful. I also want to stress that if money would be the arbiter of possibility and we wouldn't figure out whether it could be done at all, we'd be engaging in another brainless exercise of destruction, since we're going to increase demand for hard-to-get yet essential materials. Consider for instance the fact that our thirst for copper and silver trumps the existence of national forests in the US as can be seen in the case of Oak Flat / Tonto National Forest in Arizona. Have you ever wondered why this is? It is because the easy to get materials have been mined, and getting to the other less dense deposits costs more and destroys more. Solar PV wouldn't be anywhere without mines that provide copper, iron ore, Bauxite (Aluminum ore), and a plethora of rare earth materials, each of which subsequently needs chemical processing to get disassociated and purified.

Nature bats last.

Isn't it ironic that in order to build these "*life-saving energy farming devices*" we have to destroy nature in the process? And I am not just homing in on the required lands on which to construct these, or the waste sites required to store the stuff that is left after the functional lives of these technologies. We also have to reckon the amount of mining required and the sheer volume chemical waste produced on a per MW basis.

Why should we bother?

The prospects do not look good. In fact, it is questionable if we ever get *there* in time. This truly is the Apollo 11 moment for humanity. I'm not being facetious and even though this might sound like hyperbole it is perfectly justified. by the time, we will truly act and begin to shrink the combustion economy it probably will be too late, and it is going to take the full measure of our collective intelligence, ingenuity, and technological fortitude to get out of this mess. Will we be fast enough? When will the alarm bells truly ring? Which feat of destruction will scare us so much that we finally start whatever process is needed to end this madness?

Fortunately, countries like China, Russia, India, and Canada are leading us out of this combustion economy albeit very slowly and gradually and in the process also still clinging on to fossil fuels. However, they are in the process of innovating and building and exporting more nuclear energy which by any means is the only meaningful thing we can do in order to stop the influence of coal and gas and oil.

Europe is being torn apart as countries as Sweden and Germany are seriously working against nuclear energy and a new left-wing government in France also is envisioning a future without nuclear. I must say that these moves are quite paradoxical and can only be explained by the influence of popular demand, the electorate, the mandate, and the dominant ideologies that rule. It is incredibly stupid to see Sweden and Germany working to shut down their nuclear reactors while it is perfectly clear that this only leads to an increase the consumption of energy generated from biomass, coal, and lignite. This is a paradox seems to be inexplicable—or justified might be a better word for it—from an environmental viewpoint.

Fortunately, the list of influential and economically potent people convinced of the merit of nuclear energy is growing. Let's take a look at some of the prominents that already are, or may be shifting towards being pro-nuclear, rather than anti...

Excerpted from Al Gore's statements at TED in February 2016:

"if you add the projections for nuclear on here, particularly if you assume that the work many are doing to try to break through to safer and more acceptable, more affordable forms of nuclear, this could change even more dramatically."

I do not agree with all that he has said during this talk, especially when looking at certain things he presents as fact, but we may note that he is not sharing a blatant anti-nuclear message, yet one that includes it. It might be an accurate assessment if I say that I think that Al Gore is actually expecting nuclear energy to become the pervasive technology but runs with the renewable ball as long as it is perceived as the most credible—and popular?—solution.

Science a la carte

James Hansen has been on the forefront of raising climate awareness for years and now he and Ken Caldeira, and Kerry Emanuel and Tom Wrigley are raising awareness as well. They, like many others, are now making the case for an all-inclusive energy solution. I consider this to be a well reasoned and well substantiated counter move to the seemingly delusional movement of 100% WWS proponents. These seasoned scientists gave a press conference at the COP21 summit. Their stance is supported by countless of hours of research a firm grounding in physics and scientific literacy and an intimate knowledge about energy generation and consumption. These people should be heard and should be taken seriously, not because of their academic pedigree—which is considerable—but because their claims are well-substantiated by evidence and research. Pay no heed to the vain attempts of the anti-nuclear crowd to discredit these people as non-experts.

Where Elon Musk is heavily invested in electricity storage and solar power, he does leave some wriggle room for nuclear, and it is almost always included in the energy mix he foresees for the future. Fortunately, he often reiterates "*I am not against nuclear energy.*" This makes perfect sense because the electrical vehicle revolution he is ushering in is a perfect match for base load nuclear energy as has been shown earlier in this book. I would be immensely surprised if he doesn't already have a similar outlook on the future. He also wants to partake in the endeavor for space exploration, which is almost impossible without RTGs.

Which brings me to a subject where some people may disagree with me, they will say that the battery is too limited in terms of storage capacity, and we cannot build enough of them. The funny thing is that despite the limited storage the electric motor works at an efficiency that is around 90%—while the combustion engine is somewhere around 20%—so even if you carry less energy around, you can still achieve more because the efficiency of the engine and

167

this is the true measure of success, how to do more with less, efficiency and high-energy density trumps everything in the end.

What we really need is an Energiewende 2.0; a new and true Apollo 11 project; a new Manhattan project—as described in Highway to Dystopia. We really have to get our act together. We need to start replacing coal-fired power plants as quickly as possible. We could use gas-fired power plants as a temporary bridging technology, but far better would it be to embrace the principles advocated by Terrestrial Energy and Thorcon Power, and the other iterations of the MSRe i.e. FliBe Energy and Transatomic Power.

If we sum it up it would look like this: We need high-yield reactors with fuel efficiencies as high as possible; we need to be capable to build these at a high rate per year / per design—I suppose that it would be fair to expect somewhere in the order of 180GW or 1400 TWh of additions each year—where wind and solar would add only 500TWh for the same nameplate capacity. If we could reach these production capabilities, we could replace all fossil-fired electricity within 15 years from the point of commercialization. Sounds ambitious? Thorcon expects it to be possible with their designs. Granted the FliBe and Transatomic designs are somewhat more intricate and may take longer to build, but consider the fact that we are also able to keep building different designs, we could spread our options across the board: AP1000, APR1400, EPR, Terrapower's design, the PBR, prism reactors, and a great host of other viable designs are available to us. Until mass-produced MSRs become the mainstream, we have a big toolbox of safe and reliable reactors that could do the job.

China, for instance, has accepted Terrapower with open arms and is helping them get a test reactor built. Bill Gates is very involved with Terrapower. They get the chance to build their design over there and test it and possibly commercialize it. Why would

Science a la carte

Terrapower *snub* the US? The US has a very bad climate for nuclear innovation, their regulatory agencies are perfectly geared towards licensing contemporary designs—mainly generation I and II and III Pressurized Water Reactors and Light Water Reactors. If you want to try something different, they wouldn't know how to facilitate it, it simply isn't possible because of a bureaucratic quagmire.

We have ample opportunities to expand and improve upon our technological expertise, but we have to give the engineers and scientists some breathing room by helping their companies and institutes make their designs testable and their well-researched visions come to fruition. Consider the volume of intellectual power that is concentrated in and around the campuses in the US and their national laboratories. They have what it takes to speed up these processes, everything is there, we only have to figure out how to facilitate it, which probably is only a bureaucratic hurdle, nor more and no less. How different is Canada where you may make your case and it will be evaluated and you will be encouraged to progress as Terrestrial Energy is currently showing us.

By committing to innovation and progress in science and technology, we could get rid of coal and gas in terms of electricity within two or three decades, of that I am completely confident. It seems that the economics are there; at least that's what the MSR startups have calculated. Consider for instance the message driven home by Leslie Dewan from Transatomic, who has calculated that there's enough energy in spent fuel alone to last for another seven decades. Uranium at higher efficiencies will last for thousands of years and Thorium will add up to longer than human beings in their current form will exist. It will be our distant descendants that will be flying through interstellar space to visit our nearest star Proxima Centauri for the first time, but we will only make it that far if we commit to safe and cheap and abundant nuclear energy.

It's not only nuclear that is going to save the day. We also have to work on water infrastructure solutions, agricultural solutions, better healthcare, more efficient ways of building, smarter transportation practices, and strategies to educate everyone on the planet, not just the rich or those fortunate enough to have been born in the right spot at the right time; but all people.

One of the solutions that may be implemented is a carbon tax that is imposed on the well. It has been proposed earlier, if you would redistribute this tax to the people, they will decide what to do with the money, and if they choose to go for more efficient means because fossil fuels and their infrastructure become progressively more expensive, we would be edging away from fossil fuels in a natural and market-based way. This is something that has been championed by James Hansen over the last couple of years.

We also need to think about permanent carbon sequestration up until the point when we have stopped emitting greenhouse gases and have reduced carbon dioxide levels in the atmosphere below 350 parts per million, which seems to be a reasonable upper limit. We've now progressed well over the 400 parts per million mark, and this transgression will bring about instability and more severe weather for years to come unless we learn how to lower these levels again. This is a part of what we call *Geo-Engineering*.

Many people think that Carbon Sequestration will be an excuse for the fossil fuel companies to keep their business running. I agree with these people but we have created such a great debt of carbon emissions, the inertia of the climate system will ensure that even if we change today, we will have a debt to pay for decades. Also, consider that lowering carbon concentrations in the ocean and the atmosphere will take a very long time and needs our help. The only things we can do to decrease carbon concentrations in the air is to create new natural areas and to cultivate high-carbon removal plants like bamboo as a building material and more probable by technological carbon removal and permanent sequestration.

The concerns most people raise with sequestration is connected to the way we do it. Almost all proposals for carbon sequestration involve some kind of gaseous storage in cavernous areas underground. For instance, you extract oil or gas from the earth and replace it with carbon dioxide gas.

I don't think this is a particularly smart approach, and many climate and energy scientists show similar question marks. What we could and should do however is use a chemically driven process in which we carbonate water with carbon dioxide extracted from the air and pump it into porous basalt rock formations in the ground. Through this process, we can transform carbon dioxide into a solid compound that will be sequestered indefinitely without the risk of it ever escaping again. There's also the possibility of turning carbon dioxide pulled from the air into building materials, it is, however, unclear to me at this point which of the two paths is most efficient in terms of energy consumption.

The measure of these troubles forces us to play our hands soon. We've already seen the effects of water and food scarcity, these are the catalysts that instigate instability, turmoil, and even war! It will be getting worse as long as we tally, keep burning fossil fuels and don't introduce the alternatives that have the punch to knock them out of the field. In the end, our fossil-cravings (not the really cool fossilized life forms of yore, but the fossil fuels) may lead to a dying ocean, and this will make the dominos drop all the way up onto the land, where they will wreak further havoc on civilization.

We are fucking this planet up; I cannot say it any plainer than that.

The coral reefs are dying, plankton are balancing on the edge of a knife, drought and famine are spreading, and we're already seeing effects from large crop failures and increased stress on water resources. Time is running out, we're tallying unnecessarily, we're still arguing about total non-issues and are keeping ourselves from turning things around.

I am always straightforward; my terse and aggressive tone comes from anger and frustration. These are emotions that many reasonable people tend to keep to themselves during the process of writing these kinds of books. I choose not to, because this is serious business and it is no use trying to stick these feelings under the rug, in fact, it wouldn't feel right, I would be dishonest with you, and I am not like that... What would you do when the future of your kids hangs in the balance?

Get up! Get Serious! Don't surrender your critical faculties and keep demanding evidence and let's keep pushing forwards to the realization of credible solutions!

Do you think it is moral or wise to put the collective pressure of our greed on our children? We have a chance to set things right within a couple of decades, but we need to get started because we will be too late if we keep doing business the way we currently do. Maintaining the status quo in terms of energy generation is tantamount to increasing the trouble our children will have to cope with an unsettled climate leading to increased strains on water resources and food production. The status quo is part of the monstrum that is known as the free market, and as long as it is unguided and unregulated, we will see the consumption of carbon fuels without end. In matters of profitability, the planet bats last and as such our children also bat last...

The idea that we have from now until 2050 to get to zero carbon emissions is a huge challenge... huge! The clock is ticking, and humanity is having a party. We forget that there are more than a billion power sources on the world i.e. everything that burns anything. We're talking billions of small ones (we reach this number on cars and stoves alone...) and tens of thousands of large-scale power plants. It is a staggering challenge!

It's the impending heart-attack of civilization. Metaphorically speaking we're looking at an overweight middle-aged male with an insatiable thirst for sugar, and who has a high blood pressure and high cholesterol levels, a man who, despite stern warnings from his doctor, keeps eating fast food and drinking lots of alcohol. When will the pump fail? How will it manifest itself? Is it sea-level rise? The possible oblivion of all of the world's coastal cities and the economic hot-spots? Will it be the death of the world's food baskets? Or will it be the subsea blight that will dissolve Plankton and leave the bleached skeletons of the precious coral reefs behind?

We know that sea-level rise, in the end, will displace at least a billion people and affect the entire world population. Also, consider that once we have maximum sea-level rise, many sources of fresh water have depleted. This will be a major catastrophe, far worse than many people can imagine, once we say goodbye to water and food, we're in for it. What's next? Soylent Green? I don't think we will transition peacefully if this actually happens, It will be very bloody indeed. Along with these harrowing possibilities comes the loss of our cultural and historic heritage through the displacement of billions of people and the untold destruction that they will leave behind.

This is not the time to fight about irrelevant things, at the very moment, we are threatened with a catastrophe which will affect us for centuries. It's like we are arguing in our politics about the accommodations on the Titanic while there is a huge hole in the bottom of the ship, a meme that is often used nowadays. It's absolutely true, though! The music keeps playing...

Let's look at it from this angle, my antagonism to wind and solar is a response to the people who are willing enough to exclusively put the WWS eggs in our energy basket, probably knowing that they are proposing something totally and utterly ineffective in our plight against the effects of our combustion economy. By cherry picking negative facts to support their anti-nuclear stance they show their true nature, they either have been bought by elusive lobbyists and/or dubious financers, or—even worse—are truly dogmatic miscreants. Consider this most specious yet probable argument: these people are demagogues set out to gain from their mantras and the spread of fear and doubt...

These people are leading us astray in this quest for a truly sustainable future. We should employ an all-of-the-above strategy—not in terms of energy—to mitigate the damage done to the biosphere and to turn our influence around. As I've written in *Highway to Dystopia*, there is no one panacea; we need to change everything if we want to achieve anything in the distant future.

It should have become clear to you that, amongst many other things, nuclear energy is one of the technologies we need to employ in order to set things right. We need Nuclear, Hydro, Geothermal —and yes a little wind and a little solar—to become the sturdy pillars upon which we can build a sustainable future of plenty and prosperity and stability. We should employ them all albeit after having done the due diligence, after acknowledging which technology on the long run will come out on top, of which I am quite certain that it will be nuclear fission in the short run and eventually fusion in a time span of thousands of years—if we make it that far.

But first, we need to end this senseless debate that is being fueled by people like Mark Z. Jacobson and Joe Romm and Naomi Klein and Helen Caldicott and the suchlike. Why? Because the longer we tally and keep the public at odds, the steeper the curve we need to climb will become. In order to effectively mitigate our emissions and the damage we are wreaking on the biosphere and eventually—if at all possible—reverse it, we need to start now!

Science a la carte

Great starting points for learning more about the history and innovation of nuclear energy are Gordon Mcdowell's YouTube channels, there are no better! He has filmed a great host of explanatory lectures and talks that give an excellent insight into what is happening in the world of nuclear innovation, I learn a tremendous deal from them and watch them daily.

I want to see humanity jump forwards, to become a bastion of joy and wonder and excitement and discovery. I want emerging countries to become modern and see them contributing to our endeavor in their own ways. I want the religious to become secular and open minded and science- and technology-oriented. I want to see the women of the world pursue their own dreams and be the masters of their own fates. I want to see all the intellectual capacity of our species to become unlocked so that we may build a marvelous and incredibly vast library of books, a repository of our collective knowledge and creativity, a storehouse of the genius of our geniuses! The presence of humans on Earth would no longer be disruptive, but more in tune with and less dependent on nature. Having re-invented itself society would be a collective of beings taking care of one and other, of beings that get the opportunity to explore their own talents and possibilities. Energy-a-plenty is the key, we need copious amounts energy to build a healthy and foreseeing and optimistic society.

Appendix I

Open letters from scientists

I do not presume to know climate matters any better than those who study it, in fact, I acknowledge their expertise and examine and value the evidence they present and the conclusions that they derive from their research. These people have a firm grasp on what is going on in the world, and they should be considered and taken seriously. This doesn't mean that I take anything for granted, the accountability argument still has a large role to play, don't take anything at face value.

There are some open letters that are very valuable, and I want to share them with you because they convey a powerful yet very pressing message.

An Open Letter to Environmentalists on Nuclear Energy

To those influencing environmental policy but opposed to nuclear power:

As climate and energy scientists concerned with global climate change, we are writing to urge you to advocate the development and deployment of safer nuclear energy systems. We appreciate your organization's concern about global warming and your advocacy of renewable energy. But continued opposition to nuclear power threatens humanity's ability to avoid dangerous climate change.

We call on your organization to support the development and deployment of safer nuclear power systems as a practical means of addressing the climate change problem. Global demand for energy is growing rapidly and must continue to grow to provide the needs of developing economies. At the same time, the need to sharply reduce greenhouse gas emissions is becoming ever clearer. We can only increase energy supply while simultaneously reducing greenhouse gas emissions if new power plants turn away from using the atmosphere as a waste dump.

Renewables like wind and solar and biomass will certainly play roles in a future energy economy, but those energy sources cannot scale up fast enough to deliver cheap and reliable power at the scale the global economy requires. While it may be theoretically possible to stabilize the climate without nuclear power, in the real world there is no credible path to climate stabilization that does not include a substantial role for nuclear power

We understand that today's nuclear plants are far from perfect. Fortunately, passive safety systems and other advances can make new plants much safer. And modern nuclear technology can reduce proliferation risks and solve the waste disposal problem by burning current waste and using fuel more efficiently. Innovation and economies of scale can make new power plants even cheaper than existing plants. Regardless of these advantages, nuclear needs to be encouraged based on its societal benefits.

Quantitative analyses show that the risks associated with the expanded use of nuclear energy are orders of magnitude smaller than the risks associated with fossil fuels. No energy system is without downsides. We ask only that energy system decisions be based on facts, and not on emotions and biases that do not apply to 21st-century nuclear technology.

While there will be no single technological silver bullet, the time has come for those who take the threat of global warming seriously to embrace the development and deployment of safer nuclear power systems as one among several technologies that will be essential to any credible effort to develop an energy system that does not rely on using the atmosphere as a waste dump.

With the planet warming and carbon dioxide emissions rising faster than ever, we cannot afford to turn away from any technology that has the potential to displace a large fraction of our carbon emissions. Much has changed since the 1970s. The time has come for a fresh approach to nuclear power in the 21st century.

We ask you and your organization to demonstrate its real concern about risks from climate damage by calling for the development and deployment of advanced nuclear energy.

Sincerely,

Dr. Ken Caldeira, Senior Scientist, Department of Global Ecology, Carnegie Institution

Dr. Kerry Emanuel, Atmospheric Scientist, Massachusetts Institute of Technology

Dr. James Hansen, Climate Scientist, Columbia University Earth Institute

Dr. Tom Wigley, Climate Scientist, University of Adelaide and the National Center for Atmospheric Research

An Open Letter to Environmentalists:

As conservation scientists concerned with global depletion of biodiversity and the degradation of the human life-support system this entails, we, the co-signed, support the broad conclusions drawn in the article ***Key role for nuclear energy in global biodiversity conservation*** published in *Conservation Biology* (Brook & Bradshaw 2014).

Brook and Bradshaw argue that the full gamut of electricity-generation sources—including nuclear power—must be deployed to replace the burning of fossil fuels if we are to have any chance of mitigating severe climate change. They provide strong evidence for the need to accept a substantial role for advanced nuclear power systems with complete fuel recycling—as part of a range of sustainable energy technologies that also includes appropriate use of renewables, energy storage, and energy efficiency. This multi-pronged strategy for sustainable energy could also be more cost-effective and spare more land for biodiversity, as well as reduce non-carbon pollution (aerosols, heavy metals).

Given the historical antagonism towards nuclear energy amongst the environmental community, we accept that this stands as a controversial position. However, much as leading climate scientists have recently advocated the development of safe, next-generation nuclear energy systems to combat global climate change (Caldeira et al. 2013), we entreat the conservation and environmental community to weigh up the pros and cons of different energy sources using objective evidence and pragmatic trade-offs, rather than simply relying on idealistic perceptions of what is 'green'.

Although renewable energy sources like wind and solar will likely make increasing contributions to future energy production, these technology options face real-world problems of scalability, cost, material and land use, meaning that it is too risky to rely on them as the *only* alternatives to fossil fuels. Nuclear power—being by far the most compact and energy-dense of sources—could also make a major, and perhaps leading, contribution. As scientists, we declare that an evidence-based approach to future energy production is an essential component of securing biodiversity's future and cannot be ignored. It is time that conservationists make their voices heard in this policy arena.

Written by

Professor Barry W. Brook, Chair of Environmental Sustainability, University of Tasmania, *Australia*. barry.brook@utas.edu.au

Professor Corey J.A. Bradshaw, Sir Hubert Wilkins Chair of Climate Change, The Environment Institute, The University of Adelaide, *Australia*. corey.bradshaw@adelaide.edu.au

Signed by 74 scientists and academics

Appendix II

Reading and viewing recommendations

I am by no means an expert on any of the subjects I write about, that's why I call these books commentaries. There are plenty of people who are far more knowledgeable about these issues than I am and you should look them up and find out what they have to say about these subjects, you will learn a lot!

James Hansen—Earth Institute, Columbia University

Climate Science, Awareness and Solutions Program

http://www.columbia.edu/~jeh1/

"Storms of My Grandchildren"
ISBN 9781608195022

Ken Caldeira—Carnegie Institution for Science, Stanford University

Environmental science of climate, carbon, and energy

http://globalecology.stanford.edu/labs/caldeiralab/

Tom Wigley—DORA Fellow, University of Adelaide

Ecology and Environmental Science

http://www.adelaide.edu.au/directory/tom.wigley
http://thebreakthrough.org/people/profile/tom-wigley
https://www.uea.ac.uk/environmental-sciences/people/profile/t-wigley#overviewTab

Kerry Emanuel—Lorenz Center, Massachusetts Institute of Technology (MIT)

Earth, atmospheric and planetary sciences

http://eaps4.mit.edu/faculty/Emanuel/
https://paocweb.mit.edu/people/kerry-emanuel

"What We Know About Climate Change"
ISBN 978-0-262-05089-0

"Divine Wind: The History And Science Of Hurricanes"
ISBN 0-19-514941-6

"Atmospheric Convection"
ISBN 0-19-506630-8

Websites

www.pandoraspromise.com
www.thoriumenergyworld.com
www.thmsr.nl
www.gatesnotes.com
www.terrapower.com
www.transatomicpower.com
www.terrestrialenergy.com
www.thorconpower.com
www.flibe-energy.com
www.copenhagenatomics.com
www.atomicrod.com
www.actinideage.com
www.forbes.com—look for: James Conca
www.ted.com

www.YouTube.com—look for: Gordon Mcdowell
This channel has a great host of excellent videos concerning nuclear developments. Many interesting speakers like David Leblanc, Per Peterson, Kirk Sorensen, Robert Hargraves, Lars Jorgensen, Stephen Boyd, Alex Cannara, John Kutsch and many more.

www.skepticalscience.com
www.world-nuclear.org
www.whatisnuclear.com
www.x-lnt.org
www.mothersfornuclear.org
www.breakthroughinstitute.org
www.YouTube.com—look for: Bionerd23
www.ted.com—Many interesting talks on nuclear energy

Some essential Book and Google and YouTube searches

Niels Bohr
Albert Einstein
Robert Oppenheimer
Eugene Wigner
Glenn Seaborg
Alving Weinberg
Neil Degrasse Tyson
David Mackay
Robert Stone
Bill Gates [on Energy]
Sunniva Rose
Ben Heard
Simon Sinek
Leslie Dewann
Thomas Jam Pedersen

Gabriele Hegerl
Hans Joachim Schellnhuber
Jonathan Gregory
Drew Shindell
Andrei Sokolov
Alan Robock
Michael Mann
Stefan Rahmstorf
Chris Forest
Gavin Schmidt
Steward Brand
Taylor Wilson
Piers Forster
MIT Energy Initiative
Arctic Death Spiral

Appendix III

References & Backgrounds

Most of the assertions I've made have been the culmination of years of personal research. Most of which is unscientific, as said earlier, I'm a science voyeur, I love to peak over scientist' shoulders while they are working and sharing their work. Trying to make sense of the world, I read avidly, especially on the Internet, scientifically aligned websites in particular, but also opinion papers, and pseudo-scientific sites (one has to look at all sides of the same story). This list is far from complete. Take nothing at face value, investigate these matters yourself.

Ocean change & arctic methane

https://en.wikipedia.org/wiki/Photoelectric_effect
http://oceancurrents.rsmas.miami.edu/atlantic/gulf-stream.html
http://cdec.water.ca.gov/cdecapp/snowapp/sweq.action
http://www.sfgate.com/bayarea/article/Sierra-snowpack-at-91-of-normal-for-mid-February-6834655.php

https://en.wikipedia.org/wiki/Sierra_Nevada_%28U.S.%29
http://wxshift.com/climate-change/climate-indicators/ocean-acidification
https://archive.org/details/worldsinmakingev00arrhrich
https://www.youtube.com/watch?v=7so8GRCWA1k
http://chartsbin.com/view/2407
https://www.youtube.com/watch?v=gh9kDCuPuU8

The challenge

https://www.worldcoal.org/resources

http://ec.europa.eu/eurostat/statistics-explained/index.php/File:Total_greenhouse_gas_emissions_by_countries_%28including_international_aviation_and_excluding_LULUCF%29_1990_2013_%28million_tonnes_of_CO2_equivalents%29_updated.png

http://www.theguardian.com/environment/2012/nov/20/coal-plants-world-resources-institute

http://www.theage.com.au/victoria/will-victorias-desalination-plant-need-to-get-bigger-20160228-gn5k26.html

https://www.quora.com/How-many-homes-can-one-gigawatt-in-energy-capacity-provide-for

https://en.wikipedia.org/wiki/Renewable_energy_in_Iceland
http://www.world-nuclear.org/information-library/nuclear-fuel-cycle/nuclear-power-reactors/heavy-manufacturing-of-power-plants.aspx

http://www.world-nuclear.org/information-library/country-profiles/others/emerging-nuclear-energy-countries.aspx

http://www.eyewitnesstohistory.com/ford.htm
https://www.euronuclear.org/info/encyclopedia/r/reactor-pressure-vessel.htm

http://atomicpowerreview.blogspot.nl/2012/03/reactor-pressure-vessels-metallurgy-and.html

http://dailycaller.com/2016/03/24/china-literally-doubles-down-on-nuclear-power/

Good looking non-solutions

http://www.thoriumenergyworld.com/news/expectation-management-renewables

http://calgaryherald.com/business/energy/oldest-commercial-wind-farm-in-canada-headed-for-scrapyard-after-23-years

https://www.youtube.com/watch?v=w5v-yMFXvdY
http://www.environmentamerica.org/reports/ame/we-have-power
http://www.marketwatch.com/story/could-californias-massive-ivanpah-solar-power-plant-be-forced-to-go-dark-2016-03-16?mod=mw_share_facebook

http://www.greentechmedia.com/articles/read/ivanpah-solar-plant-falling-short-of-expected-electricity-production

http://breakingenergy.com/2015/06/17/ivanpah-solar-production-up-170-in-2015/

https://en.wikipedia.org/wiki/Ivanpah_Solar_Power_Facility
http://decarbonisesa.com/2016/03/19/ivanpah-solar-should-be-given-more-time/

http://www.theenergycollective.com/davidhone/2323160/solar-thermal-numbers

http://reports.climatecentral.org/pulp-fiction/1/
https://www.youtube.com/watch?v=kU6izpryqqw
http://science.sciencemag.org/content/319/5867/1235.abstract

Numbskull politics

http://www.bmwi.de/DE/Themen/Energie/Strommarkt-der-Zukunft/zahlen-fakten.html

https://en.wikipedia.org/wiki/Energy_in_Germany
https://en.wikipedia.org/wiki/Electricity_sector_in_Germany
http://www.forbes.com/sites/jamesconca/2012/08/31/germany-insane-or-just-plain-stupid/#2f83840e2fe0

http://dailycaller.com/2016/04/08/germany-to-abandon-1-1-trillion-wind-power-program-by-2019/#ixzz46eGTZ75K

http://actinideage.com/2015/12/31/the-lightbulb-moment/
http://cleantechnica.com/2014/11/16/whats-truth-germanys-ghg-emissions/
http://www.climatechangenews.com/2016/03/17/why-germanys-clean-energy-shift-is-vexing-its-neighbours/

http://strom-report.de/renewable-energy/#energy-transition-germany
http://www.tandfonline.com/doi/full/10.1080/00963402.2016.1145908?platform=hootsuite&

http://actinideage.com/2015/05/15/how-to-tell/
http://bravenewclimate.com/2015/05/05/environmental-and-health-impacts-of-a-policy-to-phase-out-nuclear-power-in-sweden/

http://www.svk.se/en/national-grid/the-control-room/
http://www.berliner-zeitung.de/wirtschaft/windenergie-die-bundesregierung-legt-bei-energiewende-den-rueckwaertsgang-ein-23846294

http://www.climatecentral.org/blogs/richard-nixon-the-environmentalist-resigned-38-years-ago-today-14776

http://www.nytimes.com/2015/12/04/opinion/republicans-climate-change-denial-denial.html?_r=0

http://nymag.com/daily/intelligencer/2015/09/whys-gop-only-science-denying-party-on-earth.html#

The worth-nothing of ideology

https://en.wikipedia.org/wiki/Vulnerability_of_nuclear_plants_to_attackhttp://www.wri.org/resources/data-visualizations/proposed-coal-fired-plants-installed-capacity-mw

https://www.worldcoal.org/installed-coal-generation-capacity-countryregion-1

http://pdf.wri.org/global_coal_risk_assessment.pdf
http://www.wri.org/publication/global-coal-risk-assessment

http://www.worldcoal.com/special-reports/13072015/Cleaning-up-the-coal-power-market-2551/

http://www.engineerlive.com/content/21600

http://www.helencaldicott.com/small-modular-reactors/

http://www.helencaldicott.com/the-impact-of-the-nuclear-crisis-on-global-health/

http://www.helencaldicott.com/australia-sleepwalks-towards-a-dangerous-nuclear-future/

http://www.helencaldicott.com/thorium/
http://atomicinsights.com/arnie-gundersen-has-inflated-his-resume-yet-frequently-claims-that-entergy-cannot-be-trusted/

http://www.theguardian.com/environment/2011/nov/21/christopher-busby-radiation-pills-fukushima

http://www.theecologist.org/blogs_and_comments/commentators/2987453/a_world_war_has_begun_break_the_silence.html

https://www.foe.co.uk/sites/default/files/downloads/policy-position-nuclear-power-75191.pdf

http://journals.plos.org/plosone/article?id=10.1371/journal.pone.0005738
http://web.stanford.edu/group/efmh/jacobson/Articles/I/CountriesWWS.pdf

http://www.theenergycollective.com/100-renewable-energy-nuclear-option-part-1/

http://www.ipcc.ch/ipccreports/tar/wg3/index.php?idp=128

The power of the force
http://www.statista.com/statistics/268154/number-of-planned-nuclear-reactors-in-various-countries/

https://en.wikipedia.org/wiki/APR-1400
http://www.powermag.com/south-korean-grid-connects-worlds-first-apr1400-nuclear-reactor/

http://nucleareeragione.org/2016/03/04/balance-sheet-of-electricity-generation-capacity-10-years-of-nuclear-power-at-a-glance/

https://whatisnuclear.com/articles/waste.html
https://www.epa.gov/radiation/radionuclide-basics-tritium#tab-3

http://www.thoriumenergyworld.com/news/when-will-europe-have-a-thorium-msr

http://www.nrc.gov/reading-rm/doc-collections/fact-sheets/tritium-radiation-fs.html

https://whatisnuclear.com/articles/waste.html

https://trainex.org/web_courses/tritium/reference_pages/tritium%20EPA.pdf

Appendix III

http://www.world-nuclear.org/information-library/non-power-nuclear-applications/radioisotopes-research/radioisotopes-in-medicine.aspx#ECSArticleLink5

http://www.cleannuclearpowersafehospitals.com/
https://en.wikipedia.org/wiki/Cobalt_therapy

http://www.brisbanetimes.com.au/comment/nuclear-medicine-comes-from-nuclear-reactors-20160225-gn3dlg.html

http://www.ansto.gov.au/AboutANSTO/MediaCentre/News/ACS048966
https://youtu.be/lbdlGVjjv2k
https://www.youtube.com/watch?v=xBmk7t5K35A
http://www.moltexenergy.com/aboutus/
https://www.haynesintl.com/pdf/h2052.pdf

http://www.magellanmetals.com/hastelloy_c22.html?gclid=CjwKEAjwuPi3BRClk8TyyMLloxgSJAAC0XsjwNsR1kP32Mk5TMgD9uS4AfXQSyCMolg1WGQRnTmcGhoCznrw_wcB

http://energyfromthorium.com/2006/07/07/how-to-throw-away-eight-years-worth-of-electricity/

http://www.world-nuclear.org/information-library/current-and-future-generation/thorium.aspx

http://www.daretothink.org/numbers-not-adjectives/how-long-will-our-supplies-of-uranium-and-thorium-last/

https://www.oecd-nea.org/ndd/pubs/2014/7209-uranium-2014.pdf

http://www.scientificamerican.com/article/how-long-will-global-uranium-deposits-last/

http://www.brucepower.com/cobalt-60-can-combat-zika-virus/

Anticipating the apologia argument

http://www.world-nuclear.org/information-library/nuclear-fuel-cycle/conversion-enrichment-and-fabrication/uranium-enrichment.aspx

http://www.timothymaloney.net/www.timothymaloney.net/Blog/Entries/2015/3/10_My_%28trititum%29_buckets_got_a_hole_in_it.html

http://sf.ites.utk.edu/utk/Play/f291f008e5414828b1a8ec16023ea0041d?catalog=eb238cab-f997-4587-9b7b-d0a0ab83420f

http://www.world-nuclear.org/information-library/current-and-future-generation/plans-for-new-reactors-worldwide.aspx

http://www.ltbridge.com/fueltechnology/fueltechnologyvalueproposition
https://www.technologyreview.com/s/601121/this-new-fuel-could-make-nuclear-power-safer-and-cheaper/#/set/id/601159/

https://en.wikipedia.org/wiki/Cost_of_electricity_by_source
http://thinkprogress.org/climate/2016/03/08/3757281/nuclear-industry-prices/

http://www.nrel.gov/analysis/tech_cost_dg.html
http://www.world-nuclear.org/information-library/safety-and-security/safety-of-plants/fukushima-accident.aspx

http://nuclearsafety.gc.ca/eng/resources/health/fukushima-health-reports.cfm

https://en.wikipedia.org/wiki/Fukushima_Daiichi_nuclear_disaster_casualties

http://www.unscear.org/unscear/en/fukushima.html
http://www-pub.iaea.org/MTCD/Publications/PDF/Pub1710-ReportByTheDG-Web.pdf

http://www.universal-sci.com/headlines/2016/3/16/kh8qlicgpg3idaf1lqxn8x83szvb5y

http://www.telegraph.co.uk/news/science/science-news/9094430/The-world-has-forgotten-the-real-victims-of-Fukushima.html#disqus_thread

http://go-nuclear.org/radiation/item/891-fukushima-and-radiation-as-a-terror-weapon-jane-orient

http://www.economist.com/blogs/graphicdetail/2016/03/daily-chart-5
http://www.who.int/mediacentre/news/releases/2005/pr38/en/index1.html

https://www.nrel.gov/analysis/sam/help/html-php/index.html?mtf_lcoe.htm

http://analysis.nuclearenergyinsider.com/terrestrial-targets-plant-costs-40-50mwh-bid-displace

http://www.westinghousenuclear.com/New-Plants/AP1000-PWR/Overview

https://en.wikipedia.org/wiki/AP1000

http://www.nytimes.com/reuters/2016/03/22/world/asia/22reuters-japan-nuclear-plutonium.html?_r=1

Urgency

Appendix III

http://www.ted.com/talks/al_gore_the_case_for_optimism_on_climate_change#t-45372

http://paulbeckwith.net/2015/12/02/james-hansen-talks-climate-change-at-cop21/

https://2paragraphs.com/2015/10/bill-gates-nuclear-ambitions-ending-poverty-safely/

http://qz.com/621385/bill-gates-predicts-a-clean-energy-breakthrough-within-15-years-will-save-the-planet/

http://www.forbes.com/sites/jamesconca/2012/06/10/energys-deathprint-a-price-always-paid/#7b8677cc49d2

http://nextbigfuture.com/2011/03/deaths-per-twh-by-energy-source.html
http://climatecolab.org/contests/2011/contest-2011-global/c/proposal/15102

http://www.wired.com/2016/03/russia-thinks-can-use-nukes-fly-mars-90-days-can-find-rubles/

https://www.youtube.com/watch?v=FdHo01xYTWQ